DIGITAL HEALTH

DIGITAL HEALTH

Mobile and Wearable Devices for Participatory Health Applications

Edited by

SHABBIR SYED-ABDUL, MD, MSc, PhD
Graduate Institute of Biomedical Informatics,
Taipei Medical University, Taipei, Taiwan

XINXIN ZHU, MD, PhD, FAMIA
Center for Biomedical Data Science, School of Medicine,
Yale University, New Haven, CT, United States

LUIS FERNANDEZ-LUQUE, PhD
Adhera Health Inc, Palo Alto, CA, United States

ELSEVIER

Elsevier
Radarweg 29, PO Box 211, 1000 AE Amsterdam, Netherlands
The Boulevard, Langford Lane, Kidlington, Oxford OX5 1GB, United Kingdom
50 Hampshire Street, 5th Floor, Cambridge, MA 02139, United States

Library of Congress Cataloging-in-Publication Data
A catalog record for this book is available from the Library of Congress

British Library Cataloguing-in-Publication Data
A catalogue record for this book is available from the British Library

ISBN: 978-0-12-820077-3

For information on all Elsevier publications visit our website
at https://www.elsevier.com/books-and-journals

Publisher: Mara Conner
Editorial Project Manager: Isabella Silva
Production Project Manager: Prem Kumar Kaliamoorthi
Cover Designer: Miles Hitchen

Typeset by TNQ Technologies

We dedicate this book to all advocates who endeavor to advance digital health by empowering consumers, making better-informed health decisions. To researchers, who are providing new options for facilitating prevention, early diagnosis of life-threatening diseases, and management of chronic conditions outside of traditional care settings. We thank them for their dedication to the transformation of digital health that ensures equality, empowerment, and patients' participation.

Contents

Contributors

Pamod Amarakoon, MBBS
Postgraduate Institute of Medicine, University of Colombo, Colombo, Sri Lanka

Jorge Cancela, PhD
F. Hoffmann-La Roche Ltd., Basel, Switzerland

Iliàs Charlafti, MSc
F. Hoffmann-La Roche Ltd., Basel, Switzerland

Kei Long Cheung, PhD
Department of Health Sciences, Brunel University London, London, United Kingdom

Mohamed-Amine Choukou, PhD
Department of Occupational Therapy, College of Rehabilitation Sciences, University of Manitoba, Winnipeg, MB, Canada

Emma A.D. Clifton, PhD
Department of Medicine, University of Cambridge, Cambridge, United Kingdom

Seya Colloud, PharmD
F. Hoffmann-La Roche Ltd., Basel, Switzerland

Hein de Vries, PhD
Professor in Health Communication at the Department of Health Promotion, CAPHRI Public Health, Maastricht University, Maastricht, The Netherlands

Vajira H.W. Dissanayake, MBBS, PhD
Faculty of Medicine, University of Colombo, Colombo, Sri Lanka

Noémie Elhadad, PhD
Data Science Institute, Columbia University, New York, NY, United States; Department of Biomedical Informatics, Columbia University, New York, NY, United States

Ipek Ensari, PhD
Data Science Institute, Columbia University, New York, NY, United States

Roshan Hewapathirana, MBBS, MSc, PhD
Faculty of Medicine, University of Colombo, Colombo, Sri Lanka

Santiago Hors-Fraile, MSc
Salumedia Labs, Seville, Spain

Robert Jarrin, JD
Adjunct Assistant Professor, Department of Emergency Medicine, George Washington University School of Medicine and Health Sciences, Washington, DC, United States; Adjunct Assistant Professor, School of Medicine - Department of Biochemistry and Molecular & Cellular Biology, Georgetown University Medical center, Washington, DC, United States

Juan José Lull, PhD
Instituto Universitario de Investigación de Aplicaciones de las Tecnologías de la, Información y de las Comunicaciones Avanzadas (ITACA), Universitat Politècnica de, València, Spain

Antonio Martínez-Millana, PhD
Instituto Universitario de Investigación de Aplicaciones de las Tecnologías de la, Información y de las Comunicaciones Avanzadas (ITACA), Universitat Politècnica de, València, Spain

Cecilia Mascolo, PhD
Department of Computer Science & Technology, University of Cambridge, Cambridge, United Kingdom

Kapil Parakh, MD, MPH, PhD
Adjunct Assistant Professor, Yale University School of Medicine, New Haven, CT, United States

Ignacio Perez-Pozuelo, MSc, MS
Department of Medicine, University of Cambridge, Cambridge, United Kingdom
The Alan Turing Institute, London, United Kingdom

Dimitris Spathis, MSc
Department of Computer Science & Technology, University of Cambridge, Cambridge, United Kingdom

Shabbir Syed-Abdul, MD, MSc, PhD
Graduate Institute of Biomedical Informatics, Taipei Medical University, Taipei, Taiwan

Vicente Traver, PhD
Instituto Universitario de Investigación de Aplicaciones de las Tecnologías de la, Información y de las Comunicaciones Avanzadas (ITACA), Universitat Politècnica de València, Spain

Mohy Uddin, MD
Executive Office, King Abdullah International Medical Research Center, King Saud bin Abdulaziz University for Health Sciences, Ministry of National Guard - Health Affairs, Riyadh, Kingdom of Saudi Arabia

Fei Wang, PhD
Department of Population Health Sciences, Weill Cornell Medicine, Cornell University, New York, NY, United States

Catherine Wu, MSc
F. Hoffmann-La Roche Ltd., Basel, Switzerland

Zhenxing Xu, PhD
Department of Population Health Sciences, Weill Cornell Medicine, Cornell University, New York, NY, United States

Bin Yu, PhD
American Air Liquide Inc New York, NY, United States

List of reviewers

Rosa Baños Universidad de Valencia, Spain
Jorge Cancela F. Hoffmann-La Roche AG, Switzerland
Kerstin Denecke Bern University of Applied Sciences, Switzerland
Paul R. DeMuro Royal Palm Companies, USA
Macarena Espinoza Universidad de Valencia, Spain
Faisal Farooq Qatar Computing Research Institute, Qatar
Elia Gabarron Norwegian Centre for E-health Research, University Hospital of North Norway, Norway
Juan M Garcia-Gomez Universidad Politécnica de Valencia, Spain
Mowafa Househ Hamad Bin Khalifa University, Qatar
Minna Isomursu University of Oulu, Finland
Raghvendra Mall Qatar Computing Research Institute, Qatar
Francisco Monteiro-Guerra University College Dublin, Ireland
Francisco J. Núñez-Benjumea Salumedia Labs, Spain
Sofia Ouhbi United Arab Emirates University, United Arabs Emirates
Josoa Palotti Qatar Computing Research Institute, Qatar
Chris Paton University of Oxford, UK
Vicente Traver Universidad Politécnica de Valencia, Spain
Nabil Zary Mohammed Bin Rashid University of Medicine and Health Sciences (MBRU), United Arab Emirates

Preface

The use of mobile and wearable devices for medical and wellness applications is becoming increasingly common nowadays. There are thousands of applications for promoting physical activity, helping patients who are living with chronic diseases such as diabetes and cancer. Furthermore, healthcare professionals have in their phones access to various clinical tools.

Despite all the possibilities, the design, development, validation, and scale-up of mHealth solutions are not trivial tasks. Many technical challenges remain; furthermore, usability and user engagement also present complex aspects that can cause great innovations to fail. As mHealth solutions are to be integrated into a diverse and complex societal context, special consideration needs to be given to ethical, legal, and cybersecurity aspects. This book aims to become a reference for stakeholders thinking about using or developing mobile and wearable solutions in the health domain. The book is designed for a multidisciplinary audience.

This book provides a toolbox for healthcare stakeholders involved in decision-making regarding the design, development, and implementation of mHealth solutions. Newcomers to the field from both a research and practice point of view can gain from this book an overview of the most important aspects of mHealth. This book provides study cases, key references, and insights from the personal experience of the authors.

CHAPTER 1

Introduction to digital health approach and mHealth applications for participatory health

Mohy Uddin, MD[1], Shabbir Syed-Abdul, MD, MSc, PhD[2]

[1]Executive Office, King Abdullah International Medical Research Center, King Saud bin Abdulaziz University for Health Sciences, Ministry of National Guard - Health Affairs, Riyadh, Kingdom of Saudi Arabia; [2]Graduate Institute of Biomedical Informatics, Taipei Medical University, Taipei, Taiwan

Contents

Due to the rising demands and pressing needs of global health, the healthcare industry has gone through constant flux and transformed over the time. From the inception of modern medicine in 18th century, healthcare has become dependent on technologies due to the periodic requirements; and in the beginning of 21st century, the integration of information and communication technologies (ICTs) in healthcare infrastructure became a sine qua non [1]. Looking back in 1990s, ICTs continued to accelerate from personal computers to eHealth, telemedicine, medicine 2.0, and health 2.0; and later in 2010s, the disruptive technologies came into sight to make an impact [1]. Rapid growth of medical knowledge, needs of informed decision-making, unavailability of informatics tools, increasing cost of healthcare, lack of medical manpower, financial unsustainability of healthcare systems, patient empowerment and democratization of care were some of the driving forces that reflected the imminent needs of healthcare paradigm shift, and implied to look into digital technologies and disruptive innovations in healthcare [2]. Digital health, in general, can be defined as "a cultural transformation of traditional healthcare" [2] and "use of information and communications technologies to improve human health, healthcare services, and wellness for individuals and across populations" [3]. By reforming and modernizing the healthcare services and systems in the recent decade, digital health has become part and

Digital Health
ISBN 978-0-12-820077-3
https://doi.org/10.1016/B978-0-12-820077-3.00001-8

parcel of the current healthcare practices and is considered as the cornerstone of the participatory / personalized health. With the advent of smartphones and sensing technologies, the concept of mobile health / mHealth / connected health emerged as another influential and promising name in digital health technologies. mHealth, which exploits mobile and wearable technology applications to support health, has played a pivotal role in facilitating the healthcare professionals in informed clinical decision-making by providing the information access at the point of care, connecting the healthcare stakeholders especially healthcare professionals and patients, management of health-related behaviors, early detection of diseases and remote patient monitoring [4–7]. The emergence of mobile, participatory and personalized sensing along with the health-related data capture has added another dimension for healthcare data aggregation and analysis [8,9]. mHealth not only holds the promise of healthcare transformation for individual patients but also has great prospects for population health and epidemiology as well [8,10]. Disruptive technologies, such as artificial intelligence (AI) applications in data mining and analytics using mobile and wearable devices, have opened new avenues in the field of digital health. The term of participatory health, boosted in early 2000s, presents a paradigm shift to improve the quality of patients' health by involving healthcare professionals and patients as the key partners; and is based on the constituents of digital health mentioned above, such as patient engagement and empowerment, patient-centered care and shared decision-making [4,11–13]. This shift emphasizes on the patients' well-being and wellness as the central focus by empowering the patients the opportunity to participate in shared decision-making based on their values and preferences, and ultimately bringing the positive care and healthy lifestyle for them. So, in this context, the digital health platforms, providing the required technologies and tools, are crucial for enabling patients' partnership and collaboration with the care providers to drive and support participatory health [4,14,15].

This book covers the theoretical aspects of digital health like methodologies, models, policies, regulations, opportunities and challenges in current healthcare practices first; and then reports the practical aspects of digital health approach using use cases, implementation, evaluation and sustainability of various mobile applications and wearable systems in the healthcare industry that can lead to participatory / personalized health for tomorrow's healthcare. In order to provide an overview of the book here, we will go through brief summaries of all chapters one by one. The next chapter

"Digital health in the era of personalized healthcare: opportunities and challenges for bringing research and patient care to a new level" by Cancela et al., provides overview, opportunities and promises of digital health in the context of personalized healthcare. Using examples of digital health technologies and advanced analytics in clinical research and care delivery pathways, it discusses various aspects through which healthcare can be personalized, and looks at different challenges, such as adoption and acceptance of digital health solutions. Moving forward towards the adoption of digital health in the society, the next chapter "Digital health regulatory and policy considerations" by Jarrin and Parakh, sheds light on the role of government policies, laws and regulations related to the implementation of digital health solutions. Using examples of various US government health agencies, it explains the development, reimbursement and adoption of digital health; and provides understanding of regulatory considerations from the beginning, i.e. novel ideas to the last stages, such as evaluation and impact of digital health products sequentially. In order to understand the underlying mechanism of health behavior and design health interventions, the next chapter "How to use the Integrated-Change Model to design digital health programs: pragmatic methodology" by Cheung et al., first explains the sociocognitive theories and determinants related to health behavior and outcomes, and then integrates and applies them using an Integrated-Change Model (I-Change Model), which is based on three phases: awareness, motivation and action for smoking cessation. It provides details about the pragmatic methodology to design tailored health programs in the form of four broad steps that include: selection of theoretical model, formulation of interventional goals and objectives, identification of salient beliefs, and designing the contents and algorithms for the digital health program. Based on the I-Change Model discussed in the previous chapter, the next chapter "Using the Integrated-Change Model to design digital health programs: the case of smoking cessation" by Cheung et al., elaborates the pragmatic methodology for designing tailored digital health program using a case of smoking cessation by describing both qualitative as well as quantitative evaluations and providing a graphical computer-tailored pro- gram outline. Looking into new avenues of computer tailoring, it explores the role of AI in comparing and analyzing the rule-based and data-driven tailoring for health behavior change. Moving further toward technologies and applications, the next chapter "Wearables, smartphones, and artificial intelligence for digital phenotyping and health" by Perez-Pozuelo et al., looks at mobile and wearable technologies along with AI in the digital health arena, and focuses on the digital phenotyping by reviewing the potentials of these technologies in epidemiology and clinical practices. It discusses the sensing and tracking features of mobile and wearable devices, large-scale multimodal data received through these devices, and how various AI applications can harness that data for human activity recognition, accurate predictive models of human behavior and ultimately personalized lifestyle development. By exploring mHealth and participatory research in the context of chronic diseases, the next chapter "mHealth For research: participatory research applications to gain disease insights" by Ensari and Elhadad, describes the recommendations and best practices of patient-centered mobile applications that can not only be used for patients' self-tracking but can also allow them to contribute their experiences. Using a case study of endometriosis, a Citizen Endo research project that aims to understand endometriosis through the direct involvement and inputs of patients, this chapter shows how patients' engagement and participatory design approaches in mHealth applications can facilitate informed decision-making, provide insights to chronic disease and lead to personalized care. Using mHealth application for genetic and chronic diseases, another chapter "Mobile health apps: the quest from laboratory to the market" by Noguera et al., describes the journey of MyCyFAPP Project, which is a comprehensive and interactive mobile solution aimed for personalized and accurate monitoring of patients with cystic fibrosis disease. In addition to providing continuous remote patients' monitoring to the healthcare providers, this solution also gives empowerment and self-management opportunities to the patients. This chapter also discusses the project-related marketing strategy, intellectual property rights, different possible business scenarios, several models for its exploitation and the product use cases along with the SWOT analysis. Continuing the mHealth potency for improving public health, another chapter "mHealth in public health sector: challenges and opportunities in low- and middle-income countries: a case study of Sri Lanka" by Amarakoon et al., looks at the mHealth applications in low and middle income countries using a case study of Sri Lanka to address problem of childhood malnutrition. It highlights the immense opportunities of mHealth applications in transforming the paper-based work to smart mobile device—enabled work providing related information, minimal data analytics and social recognition to the field healthcare workers. It also discusses the challenges associated with the sustainability and implementation of these systems, such as information systems cost, hardware and devices' maintenance, mobile Internet quality, proper planning, users' feedback and capacity building issues that must also be addressed from national to the field level. Going more deeper into mHealth determinants, the next chapter "Sustainability of mHealth solutions for healthcare system strengthening" by Choukou, draws attention to the sustainability aspects of mHealth for overall improvement of healthcare systems. In order to attain the democratization and sustainability of mHealth systems that can lead toward participatory healthcare, it provides various examples of mHealth applications and use cases for patients and healthcare providers; discusses the digital twin technology to enable interaction between patients and healthcare providers; describes the related economic, environmental, and social points of view for mHealth; and lists down the related barriers and enablers for mHealth systems. It finally suggests the interdisciplinary approach for the framework development, systems evaluation and policymaking for the success of future mHealth approaches.

References

[1] Mesko B. Health IT and digital health: the future of health technology is diverse. J Clin Trans Res 2018;3(Suppl. 3):431—4.
[2] Meskó B, et al. Digital health is a cultural transformation of traditional healthcare. mHealth 2017;3. 38-38.
[3] Kostkova P. Grand challenges in digital health. Front Public Health 2015;3(134).
[4] Coughlin S, et al. Looking to tomorrow's healthcare today: a participatory health perspective. Intern Med J 2018;48(1):92—6.
[5] Koydemir HC, Ozcan A. Wearable and implantable sensors for biomedical applications. Annu Rev Anal Chem 2018;11(1):127—46.
[6] Helbostad JL, et al. Mobile health applications to promote active and healthy ageing. Sensors 2017;17(3):622.
[7] Malwade S, et al. Mobile and wearable technologies in healthcare for the ageing population. Comput Methods Progr Biomed 2018;161:233—7.
[8] Clarke A, Steele R. Health participatory sensing networks. Mobile Inf Syst 2014;10:229—42.
[9] Tilak S. Real-world deployments of participatory sensing applications: current trends and future directions. ISRN Sensor Networks 2013;2013.
[10] O'Shea CJ, et al. Mobile health: an emerging technology with implications for global internal medicine. Intern Med J 2017;47(6):616—9.
[11] Macaulay AC. Participatory research: what is the history? Has the purpose changed? Fam Pract 2016;34(3):256—8.
[12] Denecke K, et al. Artificial intelligence for participatory health: applications, impact, and future implications. Yearb Med Inform 2019;28(01):165—73.

[13] Swan M. Health 2050: the realization of personalized medicine through crowd-sourcing, the quantified self, and the participatory biocitizen. J Personalized Med 2012;2(3):93–118.
[14] Jones M, DeRuyter F, Morris J. The digital health revolution and people with disabilities: perspective from the United States. Int J Environ Res Public Health 2020;17(2).
[15] Birnbaum F, et al. Patient engagement and the design of digital health. Acad Emerg Med 2015;22(6):754–6.

CHAPTER 2

Digital health in the era of personalized healthcare: opportunities and challenges for bringing research and patient care to a new level

Jorge Cancela, PhD, Iliàs Charlafti, MSc, Seya Colloud, PharmD, Catherine Wu, MSc
F. Hoffmann-La Roche Ltd., Basel, Switzerland

Contents

1. An introduction to the promise of the digital era

Personalized healthcare has the potential to transform the lives of patients by tailoring care to the individual, thus providing the most effective care and enabling the best outcomes faster than ever before. This concept is

Digital Health
ISBN 978-0-12-820077-3
https://doi.org/10.1016/B978-0-12-820077-3.00002-X
7

applicable throughout the patient journey, spanning the prevention, diagnosis, treatment, and monitoring of disease. The science of medicine has thus far allowed us to make substantial advances in diagnosing and treating diseases, but the complexity of human biology combined with external factors is staggering—every person is unique and, in many ways, so are their diseases. The digital revolution in healthcare provides new ways to collect, connect, and analyze data from each patient across large populations. This is enabling us to arrive at a deeper understanding of what distinguishes each individual and therefore how to treat a particular patient and his or her disease.

Digital health solutions can be instrumental in advancing the personalization of healthcare. The development of novel biomarkers collected through digital consumer-grade technologies such as wearable devices ("wearables") and mobile applications ("apps") on our smartphones offers the unprecedented opportunity for continuous data collection in a real-world setting. This is opening the door to new tools that are more sensitive to change than the current ones. Integrating data from digital health solutions with those from electronic health records (EHR) using cross-connected platforms will be key in maximizing the benefits of digital health innovations, and could ultimately directly impact clinical decision-making.

Despite the promise of the digital revolution in healthcare, the adoption of digital health solutions in routine clinical practice and by patients presents several challenges. The rapid evolution of technology and agile software development frameworks requires novel mechanisms for regulators to guarantee an adequate level of evidence and quality standards, to support claims, and to ensure patient safety in the digital health technology space. Moreover, the complexity of the healthcare ecosystem requires multiple stakeholders to partner and embrace a specific solution and its associated changes in relationships, such as those between patients and healthcare professionals, in order to drive its adoption in clinical practice. This is especially challenging in a space that has been traditionally fragmented and technically siloed. In addition, for digital health solutions to be made accessible, they should not only comply with regulatory requirements but also meet acceptable cost-effectiveness standards, overcome barriers to integration into daily clinical practice, and be incorporated in a convenient way that drives long-term engagement with patients, rather than adds to patients' burden. This chapter will discuss the existing opportunities around digital health in the context of personalized healthcare and the current challenges associated with its adoption in research and clinical practice, presenting illustrative examples along the way.

2. Opportunities for digital health in the context of personalized healthcare

2.1 Personalization of healthcare: the relationship between data, digital technologies and advanced analytics

Personalization of healthcare has the potential to enable the most effective healthcare to be made available to individual patients faster and at a lower overall cost to healthcare systems than before. By utilizing appropriate data from various sources, a holistic view of an individual's health can be obtained, and thus care can be tailored to the patient's need, maximizing outcomes more quickly (see Fig. 2.1).

The starting point for personalized patient care is a new wave of data. The types and amounts of data available in the past cannot be compared with the vast volumes of data now being generated from multiple new sources, including those that are both deep (from high-resolution tools such as comprehensive genomic profiling, digital pathology and advanced imaging) and longitudinal (such as real-world outcomes collected over the long term). The personalization of healthcare involves collecting and integrating these data, and utilizing the comprehensive datasets generated to arrive at a better understanding of diseases at different levels and ultimately solve unmet needs along the care pathway.

One component that is critical in supporting the personalization of healthcare is digital health. Digital health can be defined as the use of digital technologies to improve patient outcomes and enable better care delivery through evidence generation and services. It is important to note the

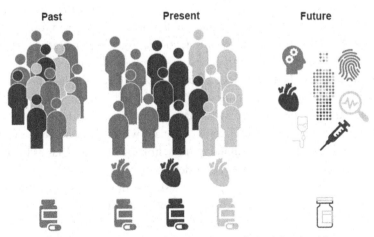

Figure 2.1 Increasing personalization is revolutionizing healthcare.

distinction between digital health tools and digital wellness tools. While digital health tools are developed on the basis that they will be used in clinical care and thus are developed with scientific rigor, are fully validated, and produce clinical-grade data, digital wellness tools are not subject to regulatory oversight. Digital health technologies can be distinguished from digital wellness tools in that they have an intended medical purpose that provides a benefit to patients; thus, they should demonstrate safety and performance in line with their intended purpose [1].

Advances in hardware are providing new sources of data, which offer the opportunity to capture clinically meaningful changes in disease at the level of the individual. While traditional clinical measures represent assessments at distinct points in time—such as cholesterol level at baseline, number of relapses per month, or an echocardiogram during an office visit—remote patient monitoring through connected devices and sensors has the potential to allow continuous data collection over the long term [2]. The evolution of mobile network capabilities and their increasing pervasiveness will facilitate continuous streaming of these data, to enable denser and richer datasets to be collected. Rather than overshadowing traditional methods of data generation, digital health technologies are providing novel insights into diseases that could complement established methods, provided the data are collated and analyzed appropriately.

Perhaps unsurprisingly, the rise in availability of large datasets has necessitated the development of innovative tools to manage, digest, visualize, and analyze these data. Equally, the availability of vast amounts of data has facilitated the creation of advanced computational tools, which require large datasets for continuous self-learning. Thus, alongside the development of new hardware, there have been parallel advances in computing techniques [2]. These tools are becoming increasingly sophisticated; artificial intelligence (AI)—based approaches (a subtype of advanced analytics including machine learning, deep learning, and natural language processing) are being developed and trained to become self-learning, and even to simulate and potentially surpass human capabilities in certain respects. In practice, advanced analytical tools will allow for automation of data extraction from clinical and digital sources, linkage of large datasets from the real world and clinical trials, and in-depth analysis of data that, given the enormity of the datasets, are impossible tasks for humans alone. These steps are required to derive meaning from complex, comprehensive datasets and transform 'raw signals' into meaningful indicators of health.

It is important to recognize that, although the digital approach to healthcare is creating a substantial amount of data, in order to have an impact on research and patient outcomes, it is not sufficient to simply have "big data." To answer scientific questions to address unmet needs, researchers need fit-for-purpose, high-quality data, derived from various sources (for example, clinical, real-world, imaging, and digital device data), carefully collated, curated, and analyzed in a manner that is most appropriate to derive scientific or clinical meaning. These meaningful data contrast with "big data" that, although available in large volumes, can be devoid of meaning owing to a lack of organization, integration, and purpose. Addressing current and future healthcare challenges will require better integration, use, and sharing of data, to allow data to become more meaningful and impactful than they currently are.

2.2 Digital health measures

Digital health measures are objective, quantifiable measures of physiology and/or behavior collected and/or measured through digital tools; in many cases, these measures form the foundation for the development of specific solutions to enable better research, clinical practice, or patient care. Digital outcome measures fall under two distinct categories—digital clinical outcomes assessments (COAs) and digital biomarkers. Although both are underpinned by digital measures, there are fundamental differences between the two (Fig. 2.2).

A COA is an objective, quantifiable measure of physiology and/or behavior used as a measure of how a patient feels, functions, or survives [3]. Clinical meaning is established de novo for COAs. Assessment of a clinical outcome can be made through report by a clinician, patient, nonclinician observer or through a performance-based assessment [3]. COAs measured through digital health technology ("digital COAs") are allowing novel measures to be developed, which have the potential to be better than traditional measures and collected remotely.

As opposed to COAs, a biomarker is a defined as a characteristic that is measured as an indicator of biological or pathological processes, or response to an intervention, including therapeutic interventions [3]. A biomarker is not an assessment of how a patient feels, functions, or survives; therefore, clinical meaning for biomarkers is established by a reliable relationship to an existing, validated endpoint [3]. Digital biomarkers such as measurements collected via sensors and combined using computational tools are becoming more accessible and are presenting opportunities for bringing clinical trials closer to patients.

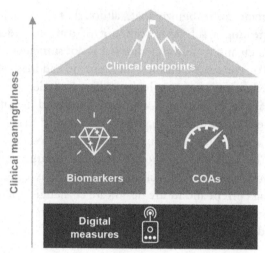

Figure 2.2 Digital endpoints are underpinned by digital biomarkers, digital clinical outcomes assessments, and digital health measures.

Biomarkers and COAs assume the highest level of clinical meaning when these measures are used as endpoints in clinical studies. Endpoints are precisely defined variables intended to reflect an outcome of interest that is statistically analyzed to address a particular research question [3]. A precise definition of an endpoint usually specifies the type of assessments made, the timing and tools used, and any other details, such as how multiple assessments from an individual will be combined [3]. Digital endpoints can be thought of as those that use digital device-generated data such as those from sensors in wearable technology to identify and/or monitor clinically relevant manifestations of health, illness, and effects of treatment. Nevertheless, it should be noted that there is not yet a widely accepted definition of digital endpoints, and views on this particular definition may differ.

Regulatory bodies are increasingly acknowledging the use of digital technologies in relation to clinical endpoints. For example, based on data provided by Sysnav's ActiMyo (Sysnav, Vernon, France) wearable device, the European Medicines Agency (EMA) recently gave a positive qualification opinion on the use of stride velocity 95th percentile as a secondary endpoint when measured at the ankle by a valid and suitable wearable device in trials of Duchenne muscular dystrophy [4]. The current pathway used in Europe to assess endpoints derived from digital health technologies is the EMA's qualification program for drug development. It involves several phases of qualification discussion with regulators and a final phase in which feedback is gathered from a broad community of experts and academics regarding the context in which the technology would be used. In the United States, the US Food and Drug Administration (FDA)'s drug development tool program has been in place for many years. This program is not often used by companies due to the lengthiness of the procedure and the low success rate. In the United States, many companies use an existing investigational new drug application in order to interact with the drug division and get more rapid feedback on any new endpoints proposed in the context of drug development. To enable endpoints derived from digital health technology to be used as a new gold standard to measure disease, adaptations of existing regulatory pathways will be needed, including mechanisms for regulators to leverage technical expertise. It is hoped that a shift toward acceptance of digitization will continue, enabling research to address previously unmet needs.

Using digital tools that span both clinical and real-world studies could provide unique opportunities for regulatory bodies and payers to make links between efficacy demonstrated in trials and real-world long-term effectiveness in clinical practice. This evidence-based approach to drug development could be used to drive value-based healthcare by improving reimbursement and access schemes. Therefore, the development of digital health tools to support validated clinical endpoints could modernize the drug development paradigm. There is an opportunity to enable smoother transitions between pre- and postmarketing data generation and, at the same time, provide better tools for physicians to help them in their therapeutic decision-making.

2.3 Examples of digital health technologies in clinical research

Digital health tools are being developed to address unmet needs pertaining to clinical trials. A clear unmet need in several diseases relates to trial efficiency. Reducing the duration of trials and the numbers of patients enrolled would enable more efficient clinical research and hence help make new treatments available more quickly. One particular therapeutic area associated with growing unmet need due to our aging population is that of dementia and age-related disorders [5]; hence, digital biomarkers in this space are relatively advanced. Early detection of changes in cognition in Alzheimer's disease (AD) in particular is becoming increasingly necessary as clinical trials move into the earlier stages of disease [5]. In this progressive condition, patients experience early but increasing cognitive decline, behavioral symptoms, and difficulties in activities of daily living [6]. Current assessments have struggled to identify meaningful measurable early signs of disease, in part owing to high variability between and within patients over time [7]. To address this need, several wearables and sensors that enable the continuous collection of multidimensional data have been developed. For example, in one study, changes in motion as assessed by a wearable ankle-mounted three-axis accelerometer were detectable in patients with AD in the absence of major clinical behavioral impairments [8]. Similarly, based on data collected via in-home passive infrared sensors, investigators were able to differentiate between individuals with mild cognitive impairment and those with normal cognition based on walking speed and its daily variability [9]. Researchers have developed a platform of sensors and devices to capture data in the home or in simulated home environments that, through wireless technologies and data analytics, can be used to assess a wide range of functions, including sleep and movement patterns, taking of medications, use of phones and computers, and opening and closing of doors [10,11]. Changes in speech patterns and semantic coherence [12,13], and in computer mouse movement patterns [14], may also be detectable in early stage AD. Collecting and collating data like these over time could allow identification of behavioral and cognitive biomarkers, which could be used to distinguish between healthy individuals and those in cognitive decline. This could enable refinement of trial eligibility criteria and efficacy endpoints and real-world clinical measures [5].

2.4 Examples of digital health technologies throughout the care delivery pathway

2.4.1 Screening and diagnosis

Along the patient pathway, each individual unmet need is providing opportunities for digital technologies to make an impact. With regard to the first step in the pathway, the introduction of digital health solutions associated with novel methods of patient engagement has the potential to widen screening programs to larger populations. This can accelerate diagnosis, which is generally associated with better patient outcomes. Accordingly, researchers are testing the possibility of screening large populations using the Apple iWatch Smartwatch (Apple Inc., Cupertino, CA, USA) [15]. Wrist-mounted wearables can measure pulse rate using photoplethysmography, with analysis of longitudinal pulse data enabling identification of potential cardiac arrhythmias such as atrial fibrillation (AF). The risks of asymptomatic AF, which include stroke, are the same as those in symptomatic patients [16,17], but can be reduced by anticoagulation [17]. Thus, it is important that patients are diagnosed and receive treatment as soon as possible. A prospective, single-arm pragmatic trial is currently ongoing that has, as of January 2018, enrolled over 400,000 participants, with the primary objective of testing the ability of the iWatch algorithm to identify pulse irregularity and variability indicative of AF [15]. This study provides an opportunity to determine the feasibility of a large-scale, completely virtual trial, which could form the basis for future studies using wearable technology to detect cardiac abnormalities in the general population.

In a similar effort to identify AF, a recent trial has assessed the ability of a self-applied wearable electrocardiogram (ECG) patch to detect AF among individuals at high risk of this arrhythmia [18]. A direct-to-participant randomized clinical trial showed that monitoring with the wearable iRhythm Zio^{XT} ECG patch (iRhythm Technologies, Inc., San Francisco, CA, USA) resulted in a higher rate of AF diagnosis within 4 months compared with routine care (3.9% vs. 0.9%) [18]. The results of this study and others investigating the iRhythm Zio^{XT} are encouraging in that arrhythmia detection rates are higher versus traditional measures [19]. Although there is a substantial risk of false positives that should be borne in mind, screening of large populations of asymptomatic individuals could accelerate diagnosis for many patients.

2.4.2 Treatment decisions, monitoring, and adherence

Once a diagnosis has been established, patients should be treated appropriately, and their disease and treatment monitored periodically. Infrequent check-ups with limited consultation times increase the risk of misreporting of changes in disease activity or treatment compliance, owing to patients' incomplete recall over long periods of time. This is especially important in chronic, progressive diseases characterized by periods of flare and remission, such as multiple sclerosis (MS) [20] and inflammatory bowel disease (IBD) [21]. Using data collected more frequently than solely during physicians' visits would enable diseases to be monitored more carefully. In the case of MS, for example, this would allow disease management to focus on slowing disease progression rather than on symptoms alone [20]. Similarly, frequent monitoring via use of electronic patient-reported outcomes (ePROs) could directly improve patient outcomes in IBD. Given that tight control of IBD through a treatment escalation algorithm has been associated with superior endoscopic outcomes versus conventional management [22], and that substantial progress in remote capturing of data in IBD has been made to enable such control [21], this is becoming a real possibility.

As well as monitoring disease, predicting how a patient might respond to a drug could save valuable time and resources. Companion diagnostics can be used to identify patients who are most likely to respond to specific therapies. For instance, the FDA-approved VENTANA anti-PD-L1 (SP142) immunohistochemical assay uses an immune cell scoring algorithm and can be used to predict which patients with cancer, such as urothelial carcinoma, are most likely to respond to treatment with atezolizumab, hence helping to target treatment to patients who will benefit most [23]. Such presentation of real-world data alongside clinical data is increasingly being expected by the authorities, for example, in the context of conditional approvals.

Sensors and digital biomarkers have also been increasingly explored with regard to response prediction in the last decade [5,24]. In Parkinson's disease, researchers have been able to predict response to dopaminergic treatment based on data obtained from typing on a computer in a pilot study [25]. Remotely collected data from unsupervised typing in a home environment were used to develop a novel algorithm, which was then used to detect and later predict participants' responses to medication as early as 21 weeks in advance, with good accuracy, in this small study [25]. Employing ecologically valid, objective data and algorithms like these could ultimately be used for adjusting treatments in clinical practice and ensuring patients receive care tailored to them as an individual.

Another example of digital health tools in relation to ongoing treatment monitoring relates to asthma. More than 40% of adults with asthma in the United States have uncontrolled asthma [26], and the frequency of use of short-acting β-agonists (SABAs) can be a proxy for asthma control [27]. In a pragmatic controlled study, use of the Propeller Health Platform—comprising an FDA-cleared sensor attached to a SABA inhaler, mobile app, predictive analytics, and healthcare provider feedback—was associated with decreased SABA use (improved disease control) versus routine care [28]. Although results could have been influenced by participants changing their behavior under observation, this example of novel information allowing patients and providers to identify triggers and emerging exacerbations could ultimately help to determine if treatment plans are working, and thus improve outcomes for patients.

Digitized continual monitoring strategies can further be useful with respect to measuring treatment adherence. In a recent open-label, long-term, observational study, adherence among 1190 pediatric patients receiving growth hormone was assessed via the easypod electronic injection device (Saizen, Merck KGaA, Darmstadt, Germany) [29]. The researchers reported significant correlations between adherence and 1-year change in height and rate of growth, based on adherence data reported directly from patients to healthcare providers via an e-Health platform [29]. The advantages of such a system are that patients with inadequate adherence and poor response to treatment can potentially be identified, enabling physicians to make decisions earlier than they would had the patient's data been collected only during office visits.

2.4.3 Patient satisfaction, quality of life, and survival

In the long term, what matters the most to patients with chronic diseases and their families is ensuring that quality of life is maximized, and the risk of early death is minimized. These meaningful outcomes can be influenced by digital health. In terms of digital management strategies, diabetes is one therapeutic area that is relatively advanced. Patients with this life-long metabolic disorder have seen a surge in availability of mobile self-management apps, and overall, these have been shown to be effective in terms of improving patients' glycemic control [30], which is associated with improved health-related quality of life [31]. One such example is the mySugr app (Roche Holding AG, Vienna, Austria), which is classified as a Class I medical device in the European Union (EU) under the medical device directive and by the FDA, and is the most widely used mobile app in diabetes [32]. The main feature of the app is that it provides patients with immediate access to certified diabetes educators when necessary based on algorithmic detection of problematic glucose patterns. This individualized approach has been successful in terms of patient satisfaction [33]. This example highlights the importance of developers accounting for the needs of individual users and stepping away from a "one-size-fits-all" approach.

In patients with cancer who are receiving chemotherapy, follow-up via patient-reported outcomes has been shown to improve health-related quality of life and survival [33]. Digitizing this strategy and using scheduled ePROs could enable timely and continuous collection of real-world symptom data that, when linked to an algorithm, could allow a rapid reaction to important safety events [33]. For example, the so-called "urgency algorithms" have been developed that automatically alert oncologists when self-scored symptoms match predefined criteria [34]. Use of ePROs and such algorithms in patients undergoing chemotherapy or follow-up for lung cancer has been associated with greater improvements in overall survival than standard follow-up methods [34,35].

2.4.4 Telehealth and digital interventions

Thus far, this chapter has focused on the ways in which digital technologies can be used to measure changes in health and disease, but there are also other important applications of digital health tools—the remote use of these tools to deliver care or even as interventions themselves. The remote delivery of care through "telehealth" offers opportunities for technology to improve access to care. Telehealth connects patients to healthcare services via videoconferencing, remote monitoring, and wireless communications. There is great scope for the use of these technologies among patient populations for whom distance and disability limit healthcare access, and their use has grown rapidly over the last 10 years [36]. Although this approach is still in its infancy, "telestroke"—whereby stroke experts are engaged remotely to aid treatment of individuals with a stroke who have not been admitted to hospital—has been particularly successful [37,38]. This is perhaps owing to a combination of factors (including a narrow therapeutic window, clinical findings visible on video, limited specialist availability) making stroke uniquely suited to telemedicine [38]. Nevertheless, there is potential for telehealth in other conditions. In low-income and rural settings, telehealth could increase capabilities at the local level through education and bridge gaps in access to care. In high-income settings, it could enable migration of care away from hospitals and integrate clinicians with diverse skill sets and reach new populations [37].

In the same vein, delivery of interventions through digital channels could directly improve clinical outcomes. Several examples in this space so far relate to psychiatric disorders. For instance, "deprexis," an evidence-based "digital treatment" comprising self-administered computerized cognitive behavioral therapy (cCBT), has been shown in metaanalyses of randomized controlled trials to be effective in improving depressive symptoms among patients with mild to moderate depression [39]. These findings add to a growing evidence base for individually tailored cCBT and mirror those seen in other disorders such as anxiety [40]. Taking the concept of digital behavioral therapy one step further by using wearable technology, an intervention deployed via Google Glass (Google Inc., Mountain View, CA, USA) and a smartphone app has been evaluated in autism spectrum disorder. The "Superpower Glass" intervention promotes facial engagement and emotion recognition and has been shown in a randomized controlled trial to improve the behavior of children with the disorder compared with the current standard of care [41].

3. Challenges throughout the life cycle of digital health solutions

Despite the rapidly growing and widespread interest in digital health, aspects related to creation, regulation, access, and reimbursement of digital health solutions present several practical challenges that limit its adoption. The increasing interest in digital science itself presents a challenge, with an increasing number of technologies being developed that are not targeted at any specific unmet need, thus simply adding noise rather than solutions. Nevertheless, regulators and health authorities are exploring the best ways to regulate and incorporate the most appropriate digital health solutions in their markets; recent milestone achievements demonstrate that digital health is moving in the right direction and being integrated into clinical practice.

3.1 Challenges in bringing digital health technologies to market

A major hurdle for any novel health technology developer is gaining appropriate approval to bring a product to market. Regulators' processes currently conflict with the rapid evolution of technology, since current regulatory requirements for digital solutions have been inherited from those for traditional medical devices that have lengthy development cycles and product life cycles. Yet, digital health solutions are rapidly proliferating, making it difficult for regulators' policies to evolve quickly enough to ensure they are appropriate for the technology, up to date, and applied consistently.

The FDA and the European Commission provide a large amount of guidance with regard to the regulation of digital technologies [42]. Because the emphasis of this chapter is on cutting-edge digital health technology, the focus here is on regulations with respect to "software as a medical device" (SaMD), which sits within the sphere of general medical devices. The International Medical Device Regulators Forum (IMDRF) provides useful guidance with respect to SaMD that can be adapted per local regulations, defining it as "software intended to be used for one or more medical purposes that perform these purposes without being part of a hardware medical device." The IMDRF recommends categorizing SaMD based on the level of impact of the information on the healthcare decision and the state of the healthcare situation or condition the SaMD is intended for (see Table 2.1) [43]. When the SaMD product can be used across

Table 2.1 Categories of software as a medical device (SaMD) as set out by the International Medical Device Regulators Forum [43].

State of healthcare situation or condition	Significance of information provided by SaMD to healthcare decision		
	Treat or diagnose	Drive clinical management	Inform clinical management
Critical	IV	III	II
Serious	III	II	I
Nonserious	II	I	I

multiple healthcare situations or conditions, the condition associated with the highest impact is used to determine the category [43].

The Center for Devices and Radiological Health (CDRH) is the center within the FDA that deals with "devices" and employs the same definition of SaMD as the IMDRF [43]. The CDRH is concerned predominantly with whether medical devices on the whole are "safe" to use and perform as intended. Thus, the CDRH assesses devices based on their intended use and indications for use (that is, the claims made about the product), and the level of risk they pose, classifying them into different levels, with increasing class number relating to increasing requirements of regulatory control necessary to provide reasonable assurance of safety and effectiveness (see Table 2.2) [44,45].

In the EU, Notified Bodies—private organizations accredited by national competent authorities—review device applications in line with the regulations developed by the European Parliament (Medical Device Regulation 2017/745 amending the Directive 93/42/EEC). Additional guidance is provided in the MEDDEV guidelines for the application of regulations developed by the Medical Device Coordination Group (MDCG) for medical devices [46]. The MDCG—which is composed of representatives of all Member States chaired by a representative of the European Commission—offers useful guidance to determine whether or not a software product should be regulated as a standalone medical device [46]. In the EU, as in the United States, medical devices are classed based on their intended use and associated level of risk (Classes I, IIa, IIb, and III). In every member state of the EU, all marketed medical devices must be CE (Conformité Européenne) marked. Class I devices that are considered low risk can be "self-certified," whereby the manufacturers themselves certify compliance with the essential requirements of the medical device directive, and depending on the claims being made, can do so without generating

Table 2.2 The US Food and Drug Administration classifies medical devices based on intended use and associate risks [44].

Class I	Class II	Class III
Devices subject to a comprehensive set of quality controls that are applicable to all classes of devices	Devices for which quality controls by themselves are insufficient to provide reasonable assurance of the safety and effectiveness of the device and for which there is sufficient information to establish special controls to provide such assurance	Devices for which quality controls by themselves are insufficient and for which there is insufficient information to establish special controls to provide reasonable assurance of the safety and effectiveness of the device; these devices pose a high risk to the patient and/or user (i.e., they sustain or support life, are implanted, or present potential unreasonable risk of illness or injury)
Premarket Notification 510(k) required, which assesses if the product demonstrates "substantial equivalence" to a predicate (an already marketed product) Most Class I and some Class II devices are exempt from 510(k) requirements, subject to certain limitations A device may be exempt if the US Food and Drug Administration determines that a 510(k) is not required to provide reasonable assurance of safety and effectiveness		*Premarket approval* required, unless device is a preamendments device (on the market prior to the passage of the medical device amendments in 1976 or substantially equivalent to such a device)
Product is "cleared" if successful		Product is "approved" if successful

clinical data [47]. Conversely, moderate- to high-risk devices (Classes IIa, IIb, and III) must undergo an external review, and clinical and/or nonclinical evidence is required to support their approval. The new Rule 11 implemented in the Medical Device Regulation 2017/745 (ANNEX VIII) will classify most types of SaMD products as Class IIa products.

To tackle the challenge of regulating ever-evolving SaMD, one new mechanism recently proposed by the FDA is the Software Precertification Program (the "Pre-Cert Program"). This program, whereby release of new software versions is made easier if a company is precertified, is currently being piloted, with several companies including Apple, Roche, and Samsung participating. With this novel approach, certain elements traditionally reviewed in a premarket submission for a SaMD product are evaluated at the organizational level in an "excellence appraisal," during which companies are evaluated based on five principles (see Fig. 2.3). Product-level evaluations then occur during the later "review determination" and "streamlined review" stages (see Fig. 2.4) [48]. If successful, the Pre-Cert Program will inform the development of a future regulatory model, which will provide streamlined, efficient regulatory oversight of software-based medical devices [48].

To date, several SaMD products and their underlying algorithms have been approved by the FDA, and the rate of approvals seems to be increasing [49]. A notable point in the history of digital health was reached in 2018 when the IDx-DR system became the first autonomous diagnostic AI system to be authorized by the FDA [50]. This system was approved for "use by healthcare providers to automatically detect more than mild diabetic retinopathy in adults (22 years of age or older) diagnosed with

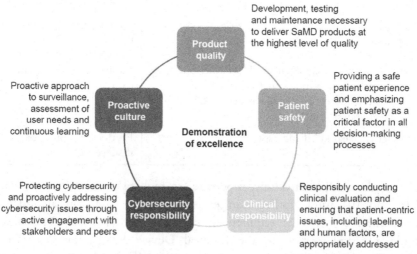

Figure 2.3 The Pre-Cert Program by the US Food and Drug Administration evaluates organizations based on five culture of quality and organizational excellence principles. *SaMD*, software as a medical device.

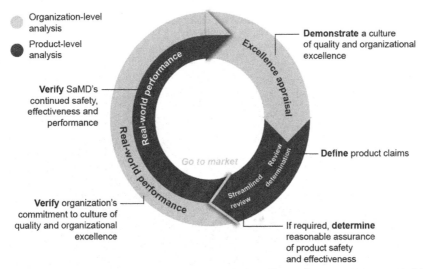

Figure 2.4 The product life cycle approach of the software Pre-Cert Program could streamline regulation of software-based medical devices [48].

diabetes who have not been previously diagnosed with diabetic retinopathy" and thus is the first approved system that provides a screening decision *without* the need for a clinician to interpret the image [50,51]. The advantages of applying AI in this manner are that the technology is useable by healthcare professionals who are not highly trained in a particular disease state, thus providing the potential for improved patient access to screening, faster and more accurate diagnosis in primary care, reduced chance of blindness, and lower overall healthcare system costs [51]. Nonetheless, there remains a paucity of publications describing most of the algorithms that have been approved. Transparency in terms of publishing details of algorithm methodology and evidence for the safety and effectiveness of AI-based technologies are greatly needed to enable researchers to overcome barriers and operate collaboratively.

3.2 Challenges in the adoption of digital health technologies

Once a SaMD product has received regulatory clearance, integration into clinical practice, access, ongoing monitoring, and management of the system present additional challenges. These challenges relate to each of the multiple stakeholders in the healthcare ecosystem—from patients and their healthcare providers to payers and reimbursement agencies—because each

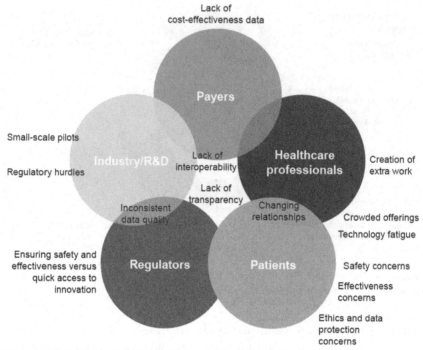

Figure 2.5 There are challenges for the adoption of digital health solutions associated with all stakeholders in the healthcare ecosystem.

stakeholder has their own incentives (see Fig. 2.5). Moreover, multiple stakeholders are required to partner to drive adoption of digital health solutions in clinical practice, and this is particularly difficult in a field that has traditionally been fragmented.

Digital health solutions should be incorporated into the healthcare system in a convenient way that drives long-term engagement with patients. As it stands, developers often focus on high-level requirements of large groups of patients, rather than the specific needs of individuals. They may develop technologies without understanding the challenges related to the disease or how a solution would fit in the patient pathway or healthcare system as a whole. Compounding this, a difficulty for patients is the paucity of accessible and reliable information on the clinical effectiveness of digital health products, combined with the ever-expanding number of solutions. It has been predicted that only a fraction of digital health solutions will be subject to a thorough regulatory review of effectiveness before being brought to market, given that products are evaluated only versus the claims they make, which may not match those that are implied [52]. Thus, the onus is on the patient to identify which solution may be helpful for them. Furthermore, even if a patient finds, for example, an app that seems effective with respect to disease management, there are several other aspects they must consider, including safety, privacy, security, interoperability, and usability, and barriers associated with these aspects may affect their likelihood of using it. A recent e-cohort study found that 35% of patients would refuse to integrate a biometric monitoring device or AI-based tool into their care, for reasons such as fears related to data misuse [53]. There is therefore a clear need for tools to assist patients in identifying appropriate technologies and making informed decisions about their use. Several initiatives have been developed with this intention; for example, the National Health Service (NHS) in the United Kingdom has created the "NHS apps library," in which it lists apps for several therapy areas that have been approved based on safety, data protection, and accessibility [54]. Although initiatives such as this represent a step in the right direction, gaps in compliance with data protection principles have been reported even with accredited health apps [55], suggesting further steps need to be taken to address these issues.

Healthcare professionals face their own barriers in the uptake of digital technologies. If a digital solution is adopted within a particular healthcare system, there is a strong chance that the relationship between patients and their healthcare professionals will change. This may not necessarily be a negative change for the patient given that healthcare in general should become more accessible. Healthcare professionals, however, may face the dilemma of whether they treat based on their experience or a clinical recommendation from an app. Furthermore, healthcare providers are increasingly required to be "tech literate," and the use of novel technologies may require training or generate more work for the individual, disincentivizing those who are not engaged. Additionally, depending on the payment model, be it fee-for-service or pay-for-performance, the way healthcare providers are reimbursed could be a factor in their readiness to adopt digital health solutions. Moreover, the healthcare system as a whole needs to be ready to act in response to demands brought about by digital health technologies, for example, if the number of people diagnosed with AF increases sharply, the system needs to ensure there is an adequate number of cardiologists available to treat and monitor these patients in the long term. This requires an understanding of clinical workflows as they operate within each hospital; data integration and system interoperability are not sufficient if patient flow and care delivery pathways cannot be reformed to adapt to the changes brought about by technologies.

The impact of changing clinical workflows and the wider long-term cost implications for health services are difficult to calculate. Although many digital health solutions are free or low cost for end users, there often remains the question of who should pay for the technology, and comprehensive economic analyses are generally lacking [56]. Accordingly, cost–benefit analyses that differ from the traditional models may be needed, for example, analyses taking patient engagement into account. These challenges feed into an overall lack of value-based business models and reimbursement schemes, which are needed for the digital health and personalized healthcare movement to be sustainable. One solution proposed by the German authorities is to have a dedicated health technology assessment and reimbursement procedure for digital health apps, in which the "quick approval" process by the BfArM (the agency responsible for drugs and devices in Germany) would assess safety, data protection, and usability. The system would allow patients to be eligible for "prescriptions" for health apps by their physician, with the price freely determined by manufacturers and reimbursed by statutory health insurance for 1 year. Thereafter, the price would be negotiated with the umbrella payer group, provided manufacturers demonstrate the "positive" effects of their app in terms of patient care [57].

As well as business strategy-related challenges, industry also faces difficulties with respect to data fragmentation. Most pilot studies have been small and siloed, and hence the wider issues of generalizability and lack of data sharing remain, making it difficult for a consensus to be reached. Cross-functional platforms and solutions are needed to enable companies to share data, progress, and success. One progressive example in this regard is the Floodlight Open study [58]. In this study, data are collected using an app that utilizes a range of sensors, including a gyroscope, accelerometer, and touch sensor, developed for use in MS, for passive measurement of mobility and active tests of hand function and cognition [59]. The technology may help identify novel digital endpoints, yet the standout feature of the approach is the open and shared stance being taken. The ongoing study is open access, meaning that anyone with MS can join. Moreover, the anonymized data collected will be freely available to clinicians and scientists, enabling different groups to conduct their own research and encouraging collaboration.

Pharmaceutical companies are also beginning to collaborate more with each other, with the industry as a whole gradually realizing the value that will come from the connection of different solutions and the creation of an overall ecosystem, similar to that seen in other sectors such as the consumer electronics industry. An example of this forward movement is that Novo Nordisk recently announced partnerships with other companies active in the diabetes space including Dexcom, Glooko, and Roche, with the vision of enabling integration of their connected insulin pen devices with broader digital health solutions. It is thought that other companies will soon follow suit, eventually reflecting a shift in mindset enabling the widespread sharing of data that is needed to overcome barriers to the creation and adoption of digital health solutions. By the same token, several EHR providers are now offering different opportunities for collaboration through partnership programs [60,61]. These approaches offer third-party solutions that leverage the existing implanted EHR technologies across healthcare settings, facilitating integration into routine clinical practice and increasing the value offering and convenience for end users.

4. Conclusions

Personalization of healthcare has the potential to enable the most effective care to be made available to individuals faster and eventually at a lower overall system cost than ever before. This new paradigm focuses not only on delivering treatment but also on achieving the best possible clinical outcomes in the long term. Digital health solutions that complement existing approaches or provide novel opportunities will play a key role in this revolution, by providing new channels for monitoring and data collection, supporting decision-making, and facilitating access to delivery of care. Nevertheless, moving from a research setting to the real world presents some unique challenges, including balancing regulatory compliance to guarantee the patient's safety and data interpretability with short software life cycles, and integrating new ways of working into an already complex multistakeholder environment. This includes a more user-centric approach that enables long-term engagement of patients with connected devices. Significant efforts have been made in terms of shifting from a traditional, fragmented approach to one that is open and collaborative. Continuing to do this will require better integration, use, and sharing of data and will ultimately result in the best possible clinical outcomes for each individual patient.

Acknowledgments

Writing support was provided by Katie Pillidge, PhD, of PharmaGenesis London, London, UK, and funded by F. Hoffmann-La Roche Ltd.

Conflicts of interest

JC is an employee of F. Hoffmann-La Roche Ltd. and a former employee of Ares Trading SA, an affiliate of Merck KGaA. IC, SC, and CW are employees of F. Hoffmann-La Roche Ltd.

References

[1] International Medical Device Regulators Forum. 2013. Available from: http://www. imdrf.org/docs/imdrf/final/technical/imdrf-tech-131209-samd-key-definitions-140 901.pdf. [Accessed February 2020].
[2] Coravos A, Goldsack JC, Karlin DR, et al. Digital medicine: a primer on measurement. Digit Biomarkers 2019;3:31–71.
[3] Food and Drug Administration. BEST (Biomarkers, EndpointS, and other tools) resource. Silver Spring (MD); 2016.
[4] European Medicines Agency. 2019. Available from: https://www.ema.europa.eu/en/ documents/scientific-guideline/qualification-opinion-stride-velocity-95th-centile-secondary-endpoint-duchenne-muscular-dystrophy_en.pdf. [Accessed February 2020].
[5] Gold M, Amatniek J, Carrillo MC, et al. Digital technologies as biomarkers, clinical outcomes assessment, and recruitment tools in Alzheimer's disease clinical trials. Alzheim Dement (N Y) 2018;4:234–42.
[6] Burns A, Iliffe S. Alzheimer's disease. BMJ 2009;338:b158.
[7] Schott JM, Aisen PS, Cummings JL, Howard RJ, Fox NC. Unsuccessful trials of therapies for Alzheimer's disease. Lancet 2019;393:29.
[8] Kirste T, Hoffmeyer A, Koldrack P, et al. Detecting the effect of Alzheimer's disease on everyday motion behavior. J Alzheim Dis 2014;38:121–32.
[9] Dodge HH, Mattek NC, Austin D, Hayes TL, Kaye JA. In-home walking speeds and variability trajectories associated with mild cognitive impairment. Neurology 2012;78:1946–52.
[10] Kaye JA, Maxwell SA, Mattek N, et al. Intelligent systems for assessing aging changes: home-based, unobtrusive, and continuous assessment of aging. J Gerontol B Psychol Sci Soc Sci 2011;66:180–90.
[11] Lyons BE, Austin D, Seelye A, et al. Pervasive computing technologies to continuously assess Alzheimer's disease progression and intervention efficacy. Front Aging Neurosci 2015;7:102.
[12] Forbes KE, Venneri A, Shanks MF. Distinct patterns of spontaneous speech deterioration: an early predictor of Alzheimer's disease. Brain Cognit 2002;48:356–61.
[13] Boschi V, Catricala E, Consonni M, Chesi C, Moro A, Cappa SF. Connected speech in neurodegenerative language disorders: a review. Front Psychol 2017;8:269.
[14] Seelye A, Hagler S, Mattek N, et al. Computer mouse movement patterns: a potential marker of mild cognitive impairment. Alzheim Dement (Amst) 2015;1:472–80.
[15] Turakhia MP, Desai M, Hedlin H, et al. Rationale and design of a large-scale, app-based study to identify cardiac arrhythmias using a smartwatch: the Apple Heart Study. Am Heart J 2019;207:66–75.

[16] Xiong Q, Proietti M, Senoo K, Lip GY. Asymptomatic versus symptomatic atrial fibrillation: A systematic review of age/gender differences and cardiovascular outcomes. Int J Cardiol 2015;191:172−7.

[17] Hart RG, Pearce LA, Aguilar MI. Meta-analysis: antithrombotic therapy to prevent stroke in patients who have nonvalvular atrial fibrillation. Ann Intern Med 2007;146:857−67.

[18] Steinhubl SR, Waalen J, Edwards AM, et al. Effect of a home-based wearable continuous ECG monitoring patch on detection of undiagnosed atrial fibrillation: the mSToPS randomized clinical trial. J Am Med Assoc 2018;320:146−55.

[19] Yenikomshian M, Jarvis J, Patton C, et al. Cardiac arrhythmia detection outcomes among patients monitored with the Zio patch system: a systematic literature review. Curr Med Res Opin 2019:1−12.

[20] Giovannoni G, Butzkueven H, Dhib-Jalbut S, et al. Brain health: time matters in multiple sclerosis. Mult Scler Relat Disord 2016;9(Suppl. 1):S5−48.

[21] Atreja A. Smarter care for patients with inflammatory bowel disease: a necessity for IBD home, value-based health care and treat-to-target strategies. Inflamm Bowel Dis 2018;24:1460−1.

[22] Colombel JF, Panaccione R, Bossuyt P, et al. Effect of tight control management on Crohn's disease (CALM): a multicentre, randomised, controlled phase 3 trial. Lancet 2018;390:2779−89.

[23] Vennapusa B, Baker B, Kowanetz M, et al. Development of a PD-L1 complementary diagnostic immunohistochemistry assay (SP142) for atezolizumab. Appl Immunohistochem Mol Morphol 2019;27:92−100.

[24] Monje MHG, Foffani G, Obeso J, Sanchez-Ferro A. New sensor and wearable technologies to aid in the diagnosis and treatment monitoring of Parkinson's disease. Annu Rev Biomed Eng 2019;21:111−43.

[25] Matarazzo M, Arroyo-Gallego T, Montero P, et al. Remote monitoring of treatment response in Parkinson's disease: the habit of typing on a computer. Mov Disord 2019;34:1488−95.

[26] Fuhlbrigge A, Reed ML, Stempel DA, Ortega HO, Fanning K, Stanford RH. The status of asthma control in the U.S. adult population. Allergy Asthma Proc 2009;30:529−33.

[27] Reddel HK, Taylor DR, Bateman ED, et al. An official American Thoracic Society/European Respiratory Society statement: asthma control and exacerbations: standardizing endpoints for clinical asthma trials and clinical practice. Am J Respir Crit Care Med 2009;180:59−99.

[28] Merchant RK, Inamdar R, Quade RC. Effectiveness of population health management using the propeller health asthma platform: a randomized clinical trial. J Allergy Clin Immunol Pract 2016;4:455−63.

[29] Koledova E, Stoyanov G, Ovbude L, Davies PSW. Adherence and long-term growth outcomes: results from the easypod connect observational study (ECOS) in paediatric patients with growth disorders. Endocr Connect 2018;7:914−23.

[30] Greenwood DA, Gee PM, Fatkin KJ, Peeples M. A systematic review of reviews evaluating technology-enabled diabetes self-management education and support. J Diabetes Sci Technol 2017;11:1015−27.

[31] Rubin RR, Peyrot M. Quality of life and diabetes. Diabetes Metab Res Rev 1999;15:205−18.

[32] Debong F, Mayer H, Kober J. Real-world assessments of mySugr mobile health app. Diabetes Technol Therapeut 2019;21:235−40.

[33] Iivanainen S, Alanko T, Peltola K, et al. ePROs in the follow-up of cancer patients treated with immune checkpoint inhibitors: a retrospective study. J Cancer Res Clin Oncol 2019;145:765−74.

[34] Denis F, Lethrosne C, Pourel N, et al. Randomized trial comparing a web-mediated follow-up with routine surveillance in lung cancer patients. J Natl Cancer Inst 2017:109.

[35] Basch E, Deal AM, Dueck AC, et al. Overall survival results of a trial assessing patient-reported outcomes for symptom monitoring during routine cancer treatment. J Am Med Assoc 2017;318:197—8.

[36] American Hospital Association. 2019. Available from: https://www.aha.org/factsheet/telehealth. [Accessed February 2020].

[37] Dorsey ER, Glidden AM, Holloway MR, Birbeck GL, Schwamm LH. Teleneurology and mobile technologies: the future of neurological care. Nat Rev Neurol 2018;14:285—97.

[38] Akbik F, Hirsch JA, Chandra RV, et al. Telestroke-the promise and the challenge. Part one: growth and current practice. J Neurointerv Surg 2017;9:357—60.

[39] Twomey C, O'Reilly G, Meyer B. Effectiveness of an individually-tailored computerised CBT programme (Deprexis) for depression: a meta-analysis. Psychiatry Res 2017;256:371—7.

[40] Berger T, Urech A, Krieger T, et al. Effects of a transdiagnostic unguided Internet intervention ("velibra") for anxiety disorders in primary care: results of a randomized controlled trial. Psychol Med 2017;47:67—80.

[41] Voss C, Schwartz J, Daniels J, et al. Effect of wearable digital intervention for improving socialization in children with autism spectrum disorder: a randomized clinical trial. JAMA Pediatr 2019;173:446—54.

[42] Food and Drug Administration. 2019. Available from: https://www.fda.gov/medical-devices/digital-health/guidances-digital-health-content. [Accessed February 2020].

[43] International Medical Device Regulators Forum. 2014. Available from: http://www.imdrf.org/docs/imdrf/final/technical/imdrf-tech-140918-samd-framework-risk-categorization-141013.pdf. [Accessed February 2020].

[44] Food and Drug Administration. 2019. Available from: https://www.fda.gov/medical-devices/overview-device-regulation/classify-your-medical-device. [Accessed February 2020].

[45] Food and Drug Administration. 2019. Available from: https://www.fda.gov/medical-devices/device-advice-comprehensive-regulatory-assistance/overview-device-regulation. [Accessed February 2020].

[46] Medical Device Coordination Group. Guidance on qualification and classification of software in regulation (EU) 2017/745 — MDR and regulation (EU) 2017/746 — IVDR. 2019.

[47] European Union. 2017. Available from: https://eur-lex.europa.eu/legal-content/EN/TXT/PDF/?uri=CELEX:02017R0745-20170505&from=EN. [Accessed February 2020].

[48] Food and Drug Administration. 2019. Available from: https://www.fda.gov/medical-devices/digital-health/digital-health-software-precertification-pre-cert-program. [Accessed February 2020].

[49] Topol EJ. High-performance medicine: the convergence of human and artificial intelligence. Nat Med 2019;25:44—56.

[50] Food and Drug Administration. 2018. Available from: https://www.fda.gov/news-events/press-announcements/fda-permits-marketing-artificial-intelligence-based-device-detect-certain-diabetes-related-eye. [Accessed February 2020].

[51] Abramoff MD, Lavin PT, Birch M, Shah N, Folk JC. Pivotal trial of an autonomous AI-based diagnostic system for detection of diabetic retinopathy in primary care offices. NPJ Digit Med 2018;1:39.

[52] Mathews SC, McShea MJ, Hanley CL, Ravitz A, Labrique AB, Cohen AB. Digital health: a path to validation. Digit Med 2019;2:38.

[53] Tran VT, Riveros C, Ravaud P. Patients' views of wearable devices and AI in healthcare: findings from the ComPaRe e-cohort. Digit Med 2019;2:53.

[54] National Health Service UK. Available from: https://www.nhs.uk/apps-library/. [Accessed February 2020].

[55] Huckvale K, Prieto JT, Tilney M, Benghozi PJ, Car J. Unaddressed privacy risks in accredited health and wellness apps: a cross-sectional systematic assessment. BMC Med 2015;13:214.

[56] Iribarren SJ, Cato K, Falzon L, Stone PW. What is the economic evidence for mHealth? A systematic review of economic evaluations of mHealth solutions. PLoS One 2017;12:e0170581.

[57] BfArM. 2019. Available from: https://www.bundesgesundheitsministerium.de/digitale-versorgung-gesetz. [Accessed February 2020].

[58] Floodlight open. 2019. Available from: https://www.floodlightopen.com/en-US/. [Accessed February 2020].

[59] Midaglia L, Mulero P, Montalban X, et al. Adherence and satisfaction of smartphone- and smartwatch-based remote active testing and passive monitoring in people with multiple sclerosis: nonrandomized interventional feasibility study. J Med Internet Res 2019;21:e14863.

[60] AllScripts. 2019. Available from: https://developer.allscripts.com/. [Accessed February 2020].

[61] eClinicalworks. 2019. Available from: https://www.eclinicalworks.com/products-services/acute-care-ehr/. [Accessed February 2020].

CHAPTER 3

Wearables, smartphones, and artificial intelligence for digital phenotyping and health

Ignacio Perez-Pozuelo, MSc, MS [1,2] Dimitris Spathis, MSc [3], Emma A.D. Clifton, PhD [1], Cecilia Mascolo, PhD [3]

[1]Department of Medicine, University of Cambridge, Cambridge, United Kingdom; [2]The Alan Turing Institute, London, United Kingdom; [3]Department of Computer Science & Technology, University of Cambridge, Cambridge, United Kingdom

Contents

1. Towards digital phenotyping

Until recently, the study of human behavior has been hindered by the ability to accurately quantify its component parts. However, technological advances in wearable devices and smartphones increasingly facilitate the collection of vast amounts of multimodal data in an unobtrusive, seamless way. In particular, the use of data generated passively by these devices enables the measurement of free-living human behavior in a scalable manner. These data can be used for digital phenotyping.

Digital Health
ISBN 978-0-12-820077-3
https://doi.org/10.1016/B978-0-12-820077-3.00003-1

Digital phenotyping can be defined as "movement-by-movement quantification of the in situ individual-level human phenotype using data from personal digital devices" [1]. This new field has already generated significant research interest across epidemiology and clinical medicine. For instance, in psychiatry, objective, multimodal, continuous quantification of behavior using individuals' own devices may result in clinically useful markers that can then be used to improve diagnostics, tailor treatment, or design new intervention models [2]. Similarly, real-time feedback paired with artificial intelligence (AI) models introduces new opportunities for health and well-being applications. For example, it may be possible to develop personalized interventional feedback generated automatically based upon physiological, environmental, and social cues from mobile and wearable devices [3].

The decreasing cost and increasing capabilities of sensors embedded in mobile and wearable devices, coupled with the proliferation of data sources from social media, environmental factors, and other sources, have yielded new concepts and techniques in the quantification of well-being, mobility, and social interaction [3]. In order for the field to progress, platforms that seek scalability and equity must be developed, enabling the establishment of shared data repositories and standardized data pipelines while fostering interdisciplinary collaborations between clinicians, patients, epidemiologists, public health researchers, and computer scientists [2]. Similarly, ubiquitous monitoring of physical behavior necessitates new regulatory frameworks and raises novel privacy considerations that safeguard the rights and freedom of users.

This chapter provides an introduction to how multimodal wearable and smartphone devices can be used to derive objective measurements of physical activity and behavior. In doing so, we provide an introduction to the field of physical activity epidemiology and the transition from questionnaire-based assessments to objective monitoring through accelerometers. We explore how mobile phones can be used to track physical and psychological behaviors. Furthermore, the role and impact of AI in this emerging field of digital phenotyping is explored.

2. Mobile health

Today, an off-the-shelf smartphone is equipped with more than a dozen sensors, including chips that measure *proximity* (how close the phone is to the user's face), *acceleration, ambient light, moisture, gyroscope, compass, barometer*

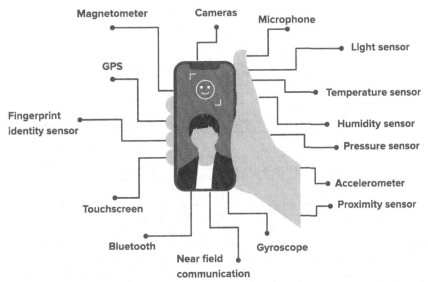

Figure 3.1 Summary of the sensors found in modern smartphones. Technological developments in smartphones enable increased processing capabilities and have equipped these mobile devices with a plethora of built-in multimodal sensors. These sensors can be used for a variety of health and wellness applications, such as mood prediction [4]. *(Figure inspired by Byrom B, Lee J, McCarthy M, Muehlhausen W. A review evaluating the validity of smartphone sensors and components to measure clinical outcomes in clinical research. Value Health 2016;19. A72.)*

(air pressure), *touch ID thumbprint*, and a *Face ID* 3D camera for secure identification (Fig. 3.1). Every phone also comes equipped with cameras, microphones, WiFi, and bluetooth connectivity as depicted in Fig. 3.2. All listed features have been used in human—computer interaction and ubiquitous computing research.

As noted in a 2010 seminal review [6], the main obstacle in mobile sensing is not that of adoption, since billions of individuals already carry sensor-rich devices. Rather, it is the ongoing challenge of performing accurate, privacy-aware research with noisy and missing data and using this research to provide effective interventions for users.

The sensors or extra metadata used in mobile health studies vary according to both the desired task and mobile phone capabilities. The most prominent inputs used to train machine learning models are presented here alongside references to the related paper in which the model was used. Movements, including specific activities, such as sitting, cycling, or walking,

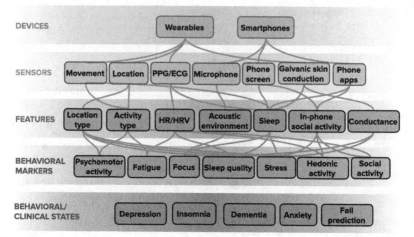

Figure 3.2 Example of a layered, hierarchical framework of wearable and mobile technology for health. The boxes at the top of the figure represent inputs to the sensing platform. The boxes in between represent features and high-level behavioral markers (*PPG*, photoplethysmography; *HRV*, heart rate variability). *(Figure inspired by Ref. Mohr DC, Zhang M, Schueller SM. Personal sensing: understanding mental health using ubiquitous sensors and machine learning. Annu Rev Clin Psychol 2017;13:23–47.)*

are estimated through the accelerometer [7–18]. Location coordinates or calculated features such as the number of places visited today can be calculated by GPS [7–10,12–16,18–21]. Ambient sounds or conversations that characterize noise are captured through the microphone [4,7–10, 12,18]. Indicators such as incoming/outgoing calls, SMSs, and emails capture communication patterns [9,12–14,16,18,19,22]. The applications used provide an indication of how users spend time on their mobile phone [9,12,18,19]. Occasional surveys related to personality, sleep quality, or current mood prompt users to provide input and act as both a ground truth to train models and as features for those models [13,14,16,18,19,22,23]. Temporal features [13,14,16,20,23], such as the day of the week or the weather [13,14,16,22], record exogenous or seasonal factors. Other task-specific features include web visits [19], phone recharge [7,18], ambient sound and light [8,9,12], keypress speed [11], compass [12], gyroscope [12], and skin physiology measures [13,14,16]. All these sensors offer new capabilities to the researchers and practitioners who build machine learning and AI models to predict clinical and behavioral outcomes. In the next section, the most widely used AI data-driven methods applied to wearable and mobile health data are discussed.

3. Artificial intelligence

Data-driven insights derived from AI have already had a tangible impact on wearable sensing and mobile health. These developments have facilitated better human activity recognition (HAR) models, more accurate predictive models of human behavior, and the development of personalized lifestyle recommendations. In this section, two schools of thought are presented regarding the application of AI methods to wearable and mobile data. The first looks for informative features that represent the time series through inventive feature extraction, while the second is based on the emerging power of representation learning to automatically extract features from lightly processed time series during the training process.

3.1 Traditional feature engineering modeling

Usually, mobile and wearable sensors data are transformed into *feature vectors* in order to be compatible with the majority of machine learning algorithms. A feature vector is a matrix-like data structure where each row represents a unique sample and each column is a separate feature or variable. However, the raw time series signals arising from, for example, accelerometers are represented as multiple continuous sequences. Consequently, the next step after data collection is to summarize the information from each sensor into a number of independent variables that capture semantic information. This task is called feature extraction and researchers work to come up with increasingly complex features that correlate with a given label. For example, the *MoodExplorer* study [12] extracted the mean, variance, and signal-to-noise ratios from microphone sensors, while the *Emotionsense* study [18] calculated the standard deviation of the magnitude of acceleration $(\sqrt{x^2 + y^2 + z^2})$ from the three axes (x, y, z) of the accelerometer.

Depending on the size of the datasets and the computing power available, computing these features as a preprocessing step can be a time-consuming, multistep process. Simple statistics such as the mean, median, standard deviation, and interquartile range are easier to estimate and could be used. However, they may not capture the informative features of noisy signals. On the other hand, higher-order statistics and transformations like the kurtosis, skewness, stationarity, least squares slope, autocorrelation, Fourier transform, and entropy provide more expressive metrics that reflect real time series phenomena like the seasonality or repeatability [24,25].

After the calculation of the appropriate metrics from the time series signal, they are then fed to machine learning algorithms. If additional linked datasets (metadata) exist (i.e., demographic or personality traits), they are concatenated with the sensor features into a big feature vector. The most common classification algorithms found in the literature are logistic regression, random forests, support vector machines, and variants of neural networks. Extensive feature extraction, resulting in a large number of features, can lead to suboptimal results. Learning algorithms underperform when the number of features is higher than the number of samples, in a phenomenon known as *("the curse of dimensionality")* [26]. As a result, researchers try to reduce the number of features before the training, either with feature selection or dimensionality reduction. For example, in a study that aimed to recognize state changes in bipolar patients [10], data were reduced using linear discriminant analysis. Other robust approaches include principal component analysis (PCA). However, it is worth noting that in a stress recognition study [22], the authors avoided PCA because the transformation yielded new variables that were hard to interpret.

3.2 Raw sensor time series modeling

The mobile sensing—ubiquitous computing research community can be compared to the computer vision community (previously known as the *image processing* community) approximately 10 years ago. A decade ago, computer vision algorithms could not work directly on the raw pixels of an image (raw sensors in our case) and researchers published inventive methods, called *feature descriptors*. Seminal papers of that time, including the scale-invariant feature transform [27], or the histogram of oriented gradients [28], are based on handcrafted algorithms that extract interest points from an image based on geometry. The turning point for computer vision took place in 2012. In that year, the *ImageNet* study [29] showed that deep learning methods can obtain better results than handcrafted-feature approaches.

The equivalent of the *ImageNet moment* has yet to arrive in mobile sensing, for a number of reasons. Foremost, datasets are not yet big enough to be fully exploited using deep learning, and there are no big, benchmark datasets that are systematically evaluated through yearly competitions. In addition, unlike object recognition, there is not a single, well-established task that guides all research in this area. As previously discussed, in this field, many overlapping but distinct aims exist (i.e., inferring mood, stress, schizophrenia, bipolar disorders, sleep patterns, social interactions, or depression).

A variety of tasks can be performed using the diverse sensors that are integrated into today's mobile phones and wearables. These sensors yield time series signals that can be modeled with recurrent or convolutional neural networks. For instance, the field of HAR has shown strong results when using deep learning methods for these tasks [30]. One of the only studies that has applied deep learning to raw time series investigated whether depressive status could be predicted by phone typing, showing 90% accuracy in depression detection based on less than a minute of typing data [11]. Phone typing dynamics is a growing area research [31]. Traditional machine learning algorithms like logistic regression or support vector machines underperformed relative to this benchmark, although the study did not perform systematic feature extraction that could have limited the potential of these techniques.

A novel, unified approach was introduced in DeepSense [17], integrating convolutional and recurrent neural networks to exploit local interactions among similar mobile sensors. This approach merged local interactions of different sensory modalities into global interactions and extracted temporal relationships to model signal dynamics. This approach demonstrates the efficacy of convolutional layers in learning local patterns and recurrent layers in learning temporal properties. The authors proposed a single network that achieves state-of-the-art results across three different problems: car tracking with motion sensors, a heterogeneous HAR task, and user identification through biometric motion analysis.

Especially in the case of time series forecasting, raw time series modeling achieves strong results when using sequence-to-sequence encoder—decoder models [32]. *Teacher forcing* training methods feed the intermediate predictions as input for multistep predictions and guarantees that it can combat the accumulated compounding errors of the predictions with adversarial networks [33]. Another research direction is to avoid recurrent and convolutional layers altogether and use only *attention* layers, since they train faster and produce cutting-edge results in some (still limited) domains [34,35].

In the next section, the application of these techniques on wearable sensors is explored in the context of epidemiological studies.

4. Toward objective measures of physical behaviors in epidemiology

4.1 Introduction to epidemiological research

The overarching goal of epidemiological research is to inform the development of interventions that reduce mortality and morbidity in populations [36]. In order to achieve this aim, epidemiologists study the distribution of health-related states or events, such as disease, in order to understand their burden and identify their determinants.[1] To conduct this type of research, intersecting data regarding both the outcome of interest and the potential determinants are required. Not only must these data intersect, with the same individuals providing information about both the exposure and the outcome, but also be both reliable and valid. This means that the measure used to assess the exposure and the outcome must be repeatable over time and accurately convey what it intends to measure. In general, objective measures are preferred, whereby individuals are not required to recall or report their exposure or outcome status themselves. This protects against unintentional recall biases and inaccuracies, as well as intentional adjustments to reporting based on social desirability. However, these considerations must also be balanced against the burden objective measuring places upon participants and the other costs that they incur. Researchers may favor a marginally less accurate measure if the measure can be collected with ease and is thus unlikely to be refused by participants, and if the cost of collection is minimal, such that many participants can be included, thus increasing the power of the study to detect associations.

Epidemiologists must also be attentive to chance, bias, confounding, and reverse causality that could cause them to draw erroneous conclusions. For example, if a study reported an association between sleep duration and obesity, the results must be interpreted with caution and cannot be assumed evidence of a causal relationship without further criteria being met. In this example, the relationship could be spurious and simply the result of chance. The probability of chance explaining the results diminishes as the number of studies reporting the same finding increase. Further, the probability of chance diminishes if larger datasets are used. If the result is not spurious, it may be the result of reverse causality. Contrary to the initial hypothesis, obesity may be the exposure variable and sleep may be the outcome.

Various methods to help rule out reverse causality exist. At minimum, longitudinal studies are required such that sleep measures are collected prior to the onset of obesity. Further, if exposures are amenable, randomized controlled trials can be conducted or if the genetic determinants of an exposure are well characterized, Mendelian randomization (MR) analyses can be performed. If reverse causality does not appear a likely explanation, it remains possible that confounding from a third, extraneous variable, associated with both the exposure and the outcome but which does not lie on

[1] https://www.who.int/topics/epidemiology/en/.

the causal pathway between them, explains the relationship [36]. For instance, smoking may cause both poor sleep and obesity, inducing a statistical association between the two variables that may erroneously be interpreted as a causal relationship. In order to control for confounding, analyses should be controlled for potential confounders or MR analyses using genetic instruments may be used.

Finally, various forms of bias should be considered. Together these comprise systematic errors in the design, conduct, or analysis of a study that may result in a distortion of the relationship between exposures and outcomes [37]. There are two major sources of bias in epidemiological research: selection bias and information bias [36]. Selection bias relates to the study population in which a research question is addressed. For example, if a study includes only adult men, the results are only generalizable to adult men and cannot be considered applicable to other population groups. A common source of selection bias in epidemiological research is that the individuals who choose enroll in population-based studies are often healthier and better educated than the general population from which they are drawn [38,39]. While the results of these studies are still internally valid, they must be generalized with caution. Biases may also relate the data collected. This is referred to as information bias and will be elaborated in the following section.

Overall, epidemiological research requires large datasets with accurate, cost-effective, and minimally burdensome measures of exposures, outcomes, and potential confounding variables. Ideally, these datasets should follow participants longitudinally. In the following sections, the way in which multimodal wearable sensing devices have revolutionized the ability to interrogate the associations of human activity to health and disease is explored.

4.2 Traditional measurement of physical activity through questionnaires

Prior to the advent of wearable sensing and mHealth technologies, researchers primarily relied upon questionnaire-based methods to measure physical activity. Questionnaires have many advantages for epidemiological research. They do not require experts or any special training to administer, and they are also cost-effective, noninvasive, and widely acceptable to participants. Further, individuals can be asked to report upon their typical, long-term habits and behaviors, which may not be accurately represented in laboratory settings. These characteristics of questionnaires facilitate the collection of data from large numbers of individuals and explains the popularity of these approaches.[2]

Despite their many advantages, questionnaires are not objective measures and may be subject to information bias. Information bias occurs when the measures used in a study are inaccurate. In the case of self-report measures, individuals may inaccurately recall their behavior and report an idealized version of their habits or some combination. Previous studies have found that self-reported physical activity suffers from reporting bias and that this results from a combination of social desirability bias (reporting behavior which is seen to be socially desirable), as well as the cognitive complexity of reporting the duration, intensity, and frequency of physical activity behaviors with precision [40–42]. In addition, the understanding of a behavior that is self-reported is limited to the specific set of questions given to study participants. These may not be enough to reflect a complete view of complex behaviors. Inaccuracies resulting from reporting errors may be randomly distributed across the population being studied. In this case, the wresults of the study would be biased toward the null, diminishing the ability of the researchers to identify true associations between exposures and outcomes. However, the errors may also be systematic, with participants in different population groups systematically under- or overreporting their activity levels. This could lead to the identification of erroneous associations.

In order to diminish concern regarding information bias in studies using self-report measures of physical activity, questionnaires should be validated against a gold standard measure.

4.3 The transition toward objective monitoring of physical behaviors

Objective monitoring of physical activity with devices such as pedometers (step measurement [43–45]), actigraphy (count-based movement measurement [46]), and accelerometers (raw movement intensity measurement) has been used to overcome the limitations of self-reported activity measures [47]. Recently, increasingly sophisticated sensors embedded within smartphones have resulted in a proliferation of *affective computing* and *behavioral phenotyping* applications, as explored in Section 2. A noncomprehensive overview of the current landscape for human behavior phenotyping using wearable sensors and smartphones is presented in Fig. 3.2.

---

[2] https://www.who.int/ncds/surveillance/steps/resources/GPAQ_Analysis_Guide.pdf.

Technological advances in the last 20 years allow for devices like triaxial accelerometers to record and store data across multiple days without requiring recharging. Further, such devices are affordable, reliable, and nonobtrusive. Indeed, in 2003, the National Institutes of Health and the National Cancer Institute funded the National Health and Nutrition Examination Survey,[3] a large epidemiological study that aims to further understand the objective measurement of physical activity through accelerometery, became the first study of its kind in the United States. Many other large initiatives followed. The UK Biobank Study,[4] the Whitehall study,[5] and the China Kadoorie Biobank[6] all exemplify the use of accelerometry in large-scale observational studies.

These studies allow researchers to perform epidemiological investigations exploring the associations between activity-related exposures of interest (predominantly comprising physical activity, sedentary behavior, and sleep) and disease outcomes, while controlling for potential confounders (such as diet, alcohol consumption, smoking habits, or socioeconomic background). Similarly, such studies often provide intersecting genome-wide genotyping information, facilitating genome-wide association studies (GWAS) designed to identify the determinants of physical activity, sedentary behavior, or sleep [48]. GWAS results can then be used to facilitate MR studies in other cohorts, designed to assess the causal impact of physical behaviors on health and disease outcomes.

4.4 Analyzing physical activity: accelerometers for movement analysis

Although physical activity and exercise are often used interchangeably in the literature, there is a difference between these concepts. Physical activity can be defined as any bodily movement that results in *energy expenditure* being increased above resting levels. Exercise is a particular type of physical activity that is purposeful, planned, structured, and often repetitive [50]. As such, activities such as housework are considered examples of physical activity, but not of exercise, because they are typically sporadic and unplanned in nature [51].

---

[3] https://www.cdc.gov/nchs/nhanes/index.htm.
[4] https://www.ukbiobank.ac.uk/.
[5] https://www.ucl.ac.uk/epidemiology-health-care/research/epidemiology-and-public-health/research/whitehall-ii.
[6] https://www.ckbiobank.org/site/.

Physical activity can be broken down and defined by (1) type (walking, running, cycling, etc.); (2) duration/volume (total time performing the activity); (3) frequency (number of sessions either per day or per week); and (4) intensity (how much energy is expended during exercise) [52]. Metabolic equivalent tasks (METs) are often used to describe the intensity of a given activity. For instance, one MET is equivalent to sitting at rest [52]. Depending on their intensity, activities can be categorized into sedentary ($\leq$ 1.5 METs), light (1.6–2.9 METs), moderate (3.0–5.9 METs), or vigorous ($\geq$ 6.0 METs) [52]. Different types of activities will normally fall into one of these buckets repeatedly. For instance, typing on a computer would be categorized as sedentary, walking is considered light, brisk walking is moderate, and running is vigorous. In order to understand physical behaviors at a population level, it is imperative to be able to accurately quantify the intensity of activities and link this to health outcomes. This informs the design physical activity recommendations, as well as the assessment of whether these recommendations are being met [51].

Accelerometry is a valuable technique for the accurate estimation of daily energy expenditure in large population studies, given its feasibility, low cost, and the existence of validation studies [53–55]. Acceleration signals are composed of a movement component, a gravitational component, and noise [56]. When conditions are static with nonrotational movement, the gravitational component is visible as the offset of one or more sensor axes and can then be used for the detection of the sensor orientation in relation to the vertical plane [56]. However, this separation is complicated when rotational movements are included as the frequency domains of the movement-related component, and the gravitational component can overlap, making it almost impossible to separate these two components using simple frequency-based filtering [57]. The inclusion of gyroscopes in addition to accelerometry helps to mitigate this problem, but they are not yet feasible for use in large-scale observational research [56,58]. A schematic of the processing and analysis of raw accelerometer signals is presented in Fig. 3.3. This process starts with raw measurements and data storage of triaxial acceleration waveforms (usually between 60 and 100 Hz), followed by a *postprocessing step* where the sensor is calibrated to local gravity, and time stamping and resampling take place. The filtering of machine noise ($\geq$ 20 Hz) follows and nonwear time is then identified. Once this *postprocessing step* finishes, summary metrics and feature extraction follow. In this step, statistical metrics and features (i.e., mean magnitude, pitch, roll, power spectra, etc.) are derived.

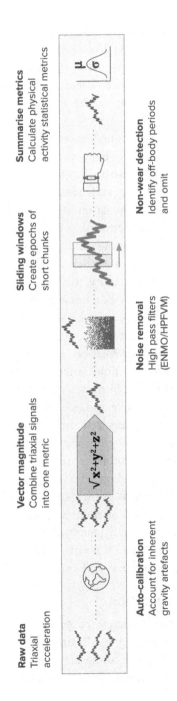

Figure 3.3 Typical data analysis pipeline for movement sensor data. From raw accelerometer data to appropriate filters and summary statistics (*ENMO*, Euclidean Norm Minus One; *HPFVM*, high-pass filtered vector magnitude).

Several accelerometer-derived metrics and constructs are well-defined, established methods to quantify objective physical activity records.

Volume of Physical Activity: Volume of physical activity refers to the total volume of activity in a given time period. In order to compare different records and recordings of different lengths, volume of physical activity is divided by the duration of the measurement to result in an average activity intensity rate.

Intensity: As previously mentioned, physical activity intensity can be categorized into vigorous, moderate, light, and sedentary. These categories were originally defined by asking participants but have since been informed by objective data, cross-referencing with resources such as the Ainsworth compendium [59], which is an aggregation of mean activity intensities that are measured or estimated while performing different activities.

Posture: Posture, limb positioning, and the pose of the body are of interest to physical activity scientists as they can provide new context for other measurements of physical activity [60,61]. Indeed, interest in this domain has grown in recent years. For instance, the consensus statement regarding the definition of sedentary behavior now includes the sitting posture as a defining characteristic [62]. Advances in micro-electromechanical sensors and orientation estimation algorithms allow wearable sensors to be used for nonrestricted human motion capture applications [63]. Biomechanically, human bodies are composed of a series of connected, jointed links that move and operate with different degrees of freedom (DOF) which can be measured using these devices. However, proper estimation of consistent and clinically meaningful joint kinematics using wearable inertial sensors requires a sensor-to-segment coordinate system calibration and understanding. To describe limb location, six parameters are required: these are location ((x,y,z) coordinates with respect to the reference system axes) and orientation parameters ((α,β,γ) angles with respect to the reference system plane) of a limb segment in space. These six coordinates constitute the DOFs of a given limb segment in space and can be used to define orientation and spatial location at a given time.

4.5 Human activity recognition

HAR can be defined as the challenge of recognizing when a person is engaging in certain activities. Hence, HAR attempts to identify the activity being performed by an individual, alongside when the activity is taking place. HAR systems are based on observations of activities that are captured using a variety of sensors, such as accelerometers and gyroscopes (to capture movement-related data), heart rate monitors (to study heart rate variability), and more. These on-body sensors allow for truly ubiquitous and continuous monitoring of physical behaviors.

The sensors record temporal data, which means that automated HAR methods face a dual issue. First, the method needs to be able to localize contiguous portions of data relevant to the activity recognition problem that the system is facing (*segmentation*). Second, those segments are then classified by automatically assigning class labels. Indeed, this task is particularly complex as information regarding the activity is typically required to identify when the activity took place. However, classification requires previous localization within the sensor dataset to determine when the activity starts and ends. Importantly, the classification step of HAR cannot retrieve any segments that are not included in the original segmentation step, making the task particularly challenging. Due to this dual problem, researchers in HAR often use *sliding-window* approaches to avoid missing any important information for the classification step. The sliding-window approach works by providing a small analysis window that shifts along the continuous data stream, extracting contiguous portions of sensor readings. The resulting data are then analyzed in isolation, showing strong results in identifying periodic activities such as walking, cycling, or climbing stairs. The performance of this analysis largely depends on how the sliding window was defined (length, steps, etc.). Thus, domain knowledge is an important factor when considering the configuration of the sliding window.

Once this process is complete, machine learning pipelines preprocess the sensor data extracted from the sliding window and proceed to extract features and employ probabilistic classification back ends that are able to assign activity labels to the corresponding analysis window [64]. Over the last decade, deep learning approaches have been established as valuable alternatives to conventional machine learning models that have limited ability in perform in the context of challenging pattern recognition tasks, such as the ones used in HAR. Deep learning models eliminate the need to manually construct feature spaces by automatically learning (*hierarchical*) data representations that are integrated into an overarching classification model. Furthermore, their modeling power has yielded very impressive results as a result of their ability to learn extremely complex decision functions. This is of great importance when dealing with the challenging analytical problems introduced in HAR tasks [65].

Combining multiple sensors for activity recognition purposes has shown promising results. These multimodal approaches have the ability to capture information that may not be possible to explore through individual sensors, such as contextual changes or social interactions [66]. In the next section, multimodal sensing is introduced, addressing the opportunities and challenges associated to integrating multiple sensors for digital phenotyping.

4.6 Multimodal sensing

Most conventional studies using either smartphones or wearable sensors to study physical behaviors have used single sensor approaches for measurement and classification tasks (accelerometer, pedometers, or gyroscopes). Occasionally they have used GPS for coarse grained location sensing. However, smartphones and new generation wearable devices often come equipped with a vast array of sensors that enable multimodal sensing. Incorporating multimodal sensing information can yield additional physiological and environmental cues, such as sound, heart rate, skin conductance, location, or activity type as depicted in the feature layer of Fig. 3.2. Indeed, large-scale longitudinal studies, such as "All of US,"[7] will incorporate multimodal wearable sensor data with the aim of better understanding physical behaviors in *free-living* environments.

Multimodal sensing approaches often rely on traditional shallow models, like random forests or support vector machines, operating on features extracted from each sensor separately [66]. Subsequently, there are two strategies to perform sensor fusion: *feature concatenation* ([67,68]) that produces a single feature vector merging all the features extracted upstream and *ensemble classifiers* ([69]) where classifiers are trained in single modalities and their predictions merged at the final step.

A significant challenge arises when attempting to incorporate information from sensor types are different in nature (i.e., an accelerometer, an ECG, and a phone camera). Due to the inherent differences in sampling rates and data distributions or shapes, the aforementioned approaches struggle to merge these diverse inputs and produce meaningful representations. An important insight here is to combine and find patterns regarding the latent cross-sensor interactions that cannot be discovered in isolation or ensembles. This is achieved with shared or merged layers in deep neural networks that can model different sensor time series and extra participant metadata in a joint latent space (see Fig. 3.4).

[7] https://allofus.nih.gov/.

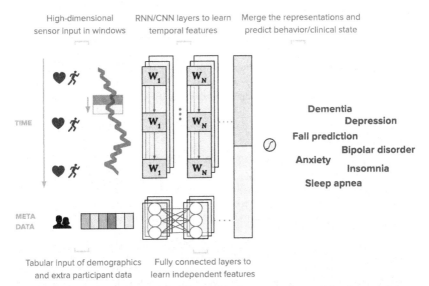

High-dimensional   RNN/CNN layers to learn   Merge the representations and
sensor input in windows   temporal features   predict behavior/clinical state

TIME

META
DATA

Tabular input of demographics   Fully connected layers to
and extra participant data   learn independent features

Figure 3.4 Multimodal sensing modeling with deep neural networks. Sensor data are modeled with time-aware layers while participant variables are fed into a separate subnetwork. The network is trained end-to-end and learns joint representations of both modalities leveraging latent combinations of sensor features and demographics (*RNN/CNN*, recurrent/convolutional neural networks).

5. Conclusion

Wearable devices and smartphones allow for truly ubiquitous and continuous tracking of physical behaviors. Here we introduced established and emerging modeling methods for mobile sensing data and discussed the impact that the application of AI will have in the field. These methods will facilitate the collection of large-scale data with unprecedented granularity which, in turn, will have important implications for industrial and academic purposes. Given the nature of the data collected, it is paramount that these practices meet appropriate privacy controls and that they are regulated accordingly. As the technology continues to develop, this will require adequate management of the availability of data for researchers to conduct studies in the public interest, while protecting personal privacy and preventing the misuse of sensitive data.

Acknowledgments

The author(s) declare(s) that there is no conflict of interest regarding the publication of this chapter. This work was supported by GSK and EPSRC through an iCase fellowship (17100053), the Embiricos Trust Scholarship of Jesus College Cambridge, and EPSRC through Grant DTP (EP/N509620/1). The icons used in some of the figures are licensed under Creative Commons by thenounproject.com.

References

[1] Torous J, Kiang MV, Lorme J, Onnela J-P. New tools for new research in psychiatry: a scalable and customizable platform to empower data driven smartphone research. JMIR mental health 2016;3:e16.

[2] Huckvale K, Venkatesh S, Christensen H. Toward clinical digital phenotyping: a timely opportunity to consider purpose, quality, and safety. NPJ Digit Med 2019;2:1−11.

[3] Woodward K, Kanjo E, Umair M, Sas C. Harnessing digital phenotyping to deliver real-time interventional bio-feedback. 2019.

[4] Spathis D, Servia-Rodriguez S, Farrahi K, Mascolo C, Rentfrow J. Passive mobile sensing and psychological traits for large scale mood prediction. In: Proceedings of the 13th EAI international Conference on pervasive computing technologies for health-care, ACM; 2019. p. 272−81.

[5] Byrom B, Lee J, McCarthy M, Muehlhausen W. A review evaluating the validity of smartphone sensors and components to measure clinical outcomes in clinical research. Value Health 2016;19. A72.

[6] Lane ND, Miluzzo E, Lu H, Peebles D, Choudhury T, Campbell AT. A survey of mobile phone sensing. IEEE Commun Mag 2010;48:140−50.

[7] Lane ND, Mohammod M, Lin M, Yang X, Lu H, Ali S, Doryab A, Berke E, Choudhury T, Campbell AT. Bewell: a smartphone application to monitor, model and promote wellbeing. In: Pervasive computing technologies for healthcare; 2011.

[8] Wang R, Chen F, Chen Z, Li T, Harari G, Tignor S, Zhou X, Ben-Zeev D, Campbell AT. Studentlife: assessing mental health, academic performance and behavioral trends of college students using smartphones. In: Proceedings of the 2014 ACM International joint Conference on pervasive and ubiquitous computing, ACM; 2014. p. 3−14.

[9] Wang R, Aung MS, Abdullah S, Brian R, Campbell AT, Choudhury T, Hauser M, Kane J, Merrill M, Scherer EA, et al. Crosscheck: toward passive sensing and detection of mental health changes in people with schizophrenia. In: Proceedings of the 2016 ACM International joint Conference on pervasive and ubiquitous computing, ACM; 2016. p. 886−97.

[10] Grünerbl A, Muaremi A, Osmani V, Bahle G, Oehler S, Tröster G, Mayora O, Haring C, Lukowicz P. Smartphone-based recognition of states and state changes in bipolar disorder patients. IEEE J Biomed Health Informat 2015;19:140−8.

[11] Cao B, Zheng L, Zhang C, Yu PS, Piscitello A, Zulueta J, Ajilore O, Ryan K, Leow AD. Deepmood: modeling mobile phone typing dynamics for mood detection. In: Proceedings of the 23rd ACM SIGKDD International Conference on knowledge discovery and data mining, ACM; 2017. p. 747−55.

[12] Zhang X, Li W, Chen X, Lu S. Moodexplorer: towards compound emotion detection via smartphone sensing. Proc. ACM Interact, Mob, Wearable Ubiquitous Technol 2018;1:176.

[13] Jaques N, Rudovic OO, Taylor S, Sano A, Picard R. Predicting tomorrow's mood, health, and stress level using personalized multitask learning and domain adaptation. Proceedings of IJCAI 2017 workshop on artificial intelligence in affective computing, vol. 66. Proceedings of Machine Learning Research; 2017. p. 17−33.

[14] Taylor SA, Jaques N, Nosakhare E, Sano A, Picard R. Personalized multitask learning for predicting tomorrow's mood, stress, and health. IEEE Trans Affect Comput 2017;99.

[15] Sano A, Amy ZY, McHill AW, Phillips AJ, Taylor S, Jaques N, Klerman EB, Picard RW. Prediction of happy-sad mood from daily behaviors and previous sleep history. In: Engineering in medicine and biology society (EMBC), 2015 37th annual International Conference of the IEEE, IEEE; 2015. p. 6796−9.

[16] Jaques N, Taylor S, Sano A, Picard R. Multimodal autoencoder: a deep learning approach to filling in missing sensor data and enabling better mood prediction. In: Proc. International Conference on affective computing and intelligent interaction (ACII), san antonio, Texas; 2017.

[17] Yao S, Hu S, Zhao Y, Zhang A, Abdelzaher T. Deepsense: a unified deep learning framework for time-series mobile sensing data processing. In: Proceedings of the 26th International Conference on world wide web, international world wide web Conferences steering committee; 2017. p. 351−60.

[18] Servia-Rodríguez S, Rachuri KK, Mascolo C, Rentfrow PJ, Lathia N, Sandstrom GM. Mobile sensing at the service of mental well-being: a large-scale longitudinal study. In: Proceedings of the 26th International Conference on world wide web, international world wide web Conferences steering committee; 2017. p. 103−12.

[19] LiKamWa R, Liu Y, Lane ND, Zhong L. Moodscope: building a mood sensor from smartphone usage patterns. In: MobiSys '13, ACM; 2013.

[20] Mikelsons G, Smith M, Mehrotra A, Musolesi M. Towards deep learning models for psychological state prediction using smartphone data: challenges and opportunities. In: In workshop on machine learning for health (ML4H) at NIPS 2017; 2017.

[21] Canzian L, Musolesi M. Trajectories of depression: unobtrusive monitoring of depressive states by means of smartphone mobility traces analysis. In: Proceedings of the 2015 ACM international joint Conference on pervasive and ubiquitous computing, ACM; 2015. p. 1293−304.

[22] Bogomolov A, Lepri B, Ferron M, Pianesi F, Pentland AS. Daily stress recognition from mobile phone data, weather conditions and individual traits. In: Proceedings of the 22nd ACM International Conference on Multimedia, ACM; 2014. p. 477−86.

[23] Suhara Y, Xu Y, Pentland A. Deepmood: forecasting depressed mood based on self-reported histories via recurrent neural networks. In: Proceedings of the 26th International Conference on world wide web, international world wide web Conferences steering committee; 2017. p. 715−24.

[24] Fulcher BD. Feature-based time-series analysis. In: Feature engineering for machine learning and data analytics, CRC press; 2018. p. 87−116.

[25] Fulcher BD, Jones NS. Highly comparative feature-based time-series classification. IEEE Trans Knowl Data Eng 2014;26:3026−37.

[26] Friedman J, Hastie T, Tibshirani R. The elements of statistical learning, vol. 1. New York: Springer series in statistics; 2001.

[27] Lowe DG. Distinctive image features from scale-invariant keypoints. Int J Comput Vis 2004;60:91−110.

[28] Dalal N, Triggs B. Histograms of oriented gradients for human detection. Computer vision and pattern recognition, 2005. CVPR 2005. IEEE computer society Conference on, vol. 1. IEEE; 2005. p. 886−93.

[29] Krizhevsky A, Sutskever I, Hinton GE. Imagenet classification with deep convolutional neural networks. In: Advances in neural information processing systems; 2012. p. 1097–105.

[30] Hammerla NY, Halloran S, Plötz T. Deep, convolutional, and recurrent models for human activity recognition using wearables. In: Proceedings of the twenty-fifth International joint Conference on artificial intelligence, AAAI press; 2016. p. 1533–40.

[31] Iakovakis D, Hadjidimitriou S, Charisis V, Bostantzopoulou S, Katsarou Z, Hadjileontiadis LJ. Touchscreen typing-pattern analysis for detecting fine motor skills decline in early-stage Parkinson's disease. Sci Rep 2018;8:1–13.

[32] Spathis D, Servia-Rodriguez S, Farrahi K, Mascolo C, Rentfrow J. Sequence multi-task learning to forecast mental wellbeing from sparse self-reported data. In: Proceedings of the 25th ACM SIGKDD International Conference on knowledge discovery & data mining, KDD '19, ACM, New York, NY, USA; 2019. p. 2886–94. https://doi.org/10.1145/3292500.3330730. http://doi.acm.org/10.1145/3292500.3330730.

[33] Lamb AM, GOYAL AGAP, Zhang Y, Zhang S, Courville AC, Bengio Y. Professor forcing: a new algorithm for training recurrent networks. In: Advances in neural information processing systems; 2016. p. 4601–9.

[34] Vaswani A, Shazeer N, Parmar N, Uszkoreit J, Jones L, Gomez AN, Kaiser Ł, Polosukhin I. Attention is all you need. In: Advances in neural information processing systems; 2017. p. 6000–10.

[35] Song H, Rajan D, Thiagarajan JJ, Spanias A. Attend and diagnose: clinical time series analysis using attention models. In: Thirty-second AAAI Conference on artificial intelligence; 2018.

[36] Gordis L. Epidemiology, A student consult title. Elsevier/Saunders; 2009. https://books.google.co.uk/books?id=GseHgIbJo4gC.

[37] Porta M. A dictionary of epidemiology. Rev Esp Salud Publica 2008;82. https://doi.org/10.1590/s1135-57272008000400008.

[38] Froom P, Melamed S, Kristal-Boneh E, Benbassat J, Ribak J. Healthy volunteer effect in industrial workers. J Clin Epidemiol 1999;52:731–5.

[39] Fry A, Littlejohns TJ, Sudlow C, Doherty N, Adamska L, Sprosen T, Collins R, Allen NE. Comparison of sociodemographic and health-related characteristics of UK biobank participants with those of the general population. Am J Epidemiol 2017;186:1026–34.

[40] Troiano RP, Berrigan D, Dodd KW, Mâsse LC, Tilert T, Mcdowell M. Physical activity in the United States measured by accelerometer. Med Sci Sports Exerc 2008;40:181–8. https://doi.org/10.1249/mss.0b013e31815a51b3.

[41] E. de Leeuw, N. Borgers, A. Smits. Pretesting questionnaires for children and adolescents, in: Methods for testing and evaluating survey questionnaires, John Wiley & Sons, Inc., Hoboken, NJ, USA, ????, pp. 409–429. http://doi.wiley.com/10.1002/0471654728.ch20. https://doi.wiley.com/10.1002/0471654728.ch20.

[42] Sallis JF, Saelens BE. Assessment of physical activity by self-report: status, limitations, and future directions. Res Q Exerc Sport 2000;71:1–14. https://doi.org/10.1080/02701367.2000.11082780.

[43] Bassett DR, Wyatt HR, Thompson H, Peters JC, Hill JO. Pedometer-measured physical activity and health behaviors in U.S. adults. Med Sci Sports Exerc 2010;42:1819–25. https://doi.org/10.1249/MSS.0b013e3181dc2e54.

[44] Corder K, Brage S, Ekelund U. Accelerometers and pedometers: methodology and clinical application. Curr Opin Clin Nutr 2007. https://doi.org/10.1097/MCO.0b013e328285d883.

[45] Schmidt MD, Blizzard LC, Venn AJ, Cochrane JA, Dwyer T. Practical considerations when using pedometers to assess physical activity in population studies: lessons from the burnie take heart study. Res Q Exerc Sport 2007;78:162—70. https://doi.org/10.1080/02701367.2007.10599413.

[46] Buchman AS, Boyle PA, Yu L, Shah RC, Wilson RS, Bennett DA. Total daily physical activity and the risk of AD and cognitive decline in older adults. Neurology 2012;78:1323—9. https://doi.org/10.1212/WNL.0b013e3182535d35.

[47] Guo W, Key TJ, Reeves GK. Accelerometer compared with questionnaire measures of physical activity in relation to body size and composition: a large cross-sectional analysis of UK Biobank. BMJ Open 2019;9. https://doi.org/10.1136/bmjopen-2018-024206.

[48] Willetts M, Hollowell S, Aslett L, Holmes C, Doherty A. Statistical machine learning of sleep and physical activity phenotypes from sensor data in 96,220 UK Biobank participants. Sci Rep 2018;8. https://doi.org/10.1038/s41598-018-26174-1.

[49] Mohr DC, Zhang M, Schueller SM. Personal sensing: understanding mental health using ubiquitous sensors and machine learning. Annu Rev Clin Psychol 2017;13:23—47.

[50] Conn VS, Hafdahl AR, Brown SA, Brown LM. Meta-analysis of patient education interventions to increase physical activity among chronically ill adults. Patient Educ Couns 2008. https://doi.org/10.1016/j.pec.2007.10.004.

[51] McCarthy M, Grey M. Motion sensor use for physical activity data: methodological considerations. Nurs Res 2015. https://doi.org/10.1097/NNR.0000000000000098.

[52] Strath SJ, Kaminsky LA, Ainsworth BE, Ekelund U, Freedson PS, Gary RA, Richardson CR, Smith DT, Swartz AM. Guide to the assessment of physical activity: clinical and research applications: a scientific statement from the American Heart association. Circulation 2013;128:2259—79. https://doi.org/10.1161/01.cir.0000435708.67487.da.

[53] Doherty A, Jackson D, Hammerla N, Plötz T, Olivier P, Granat MH, White T, van Hees VT, Trenell MI, Owen CG, Preece SJ, Gillions R, Sheard S, Peakman T, Brage S, Wareham NJ. Large scale population assessment of physical activity using wrist worn accelerometers: the UK biobank study. PloS One 2017;12:e0169649. https://doi.org/10.1371/journal.pone.0169649. http://dx.plos.org/10.1371/journal.pone.0169649.

[54] German National Cohort (GNC) Consortium. The German National Cohort: aims, study design and organization. Eur J Epidemiol 2014;29:371—82. https://doi.org/10.1007/s10654-014-9890-7. http://www.ncbi.nlm.nih.gov/pubmed/24840228.http://www.pubmedcentral.nih.gov/articlerender.fcgi?artid=PMC4050302.http://link.springer.com/10.1007/s10654-014-9890-7.

[55] Wijndaele K, Westgate K, Stephens SK, Blair SN, Bull FC, Chastin SFM, Dunstan DW, Ekelund U, Esliger DW, Freedson PS, Granat MH, Matthews CE, Owen N, Rowlands AV, Sherar LB, Tremblay MS, Troiano RP, Brage S, Healy GN. Utilization and harmonization of adult accelerometry data: review and expert consensus. Med Sci Sports Exerc 2015;47:2129—39. https://doi.org/10.1249/MSS.0000000000000661. http://www.ncbi.nlm.nih.gov/pubmed/25785929. http://www.pubmedcentral.nih.gov/articlerender.fcgi?artid=PMC4731236.

[56] Veltink P, Bussmann H, de Vries W, Martens W, Van Lummel R. Detection of static and dynamic activities using uniaxial accelerometers. IEEE Trans Rehabil Eng 1996;4:375—85. https://doi.org/10.1109/86.547939. http://ieeexplore.ieee.org/document/547939/.

[57] van Hees VT, Gorzelniak L, Dean León EC, Eder M, Pias M, Taherian S, Ekelund U, Renström F, Franks PW, Horsch A, Brage S. Separating movement and gravity components in an acceleration signal and implications for the assessment of human daily physical activity. PloS One 2013;8:e61691. https://doi.org/10.1371/journal.-pone.0061691. http://dx.plos.org/10.1371/journal.pone.0061691.

[58] Sabatini A. Quaternion-based extended kalman filter for determining orientation by inertial and magnetic sensing. IEEE (Inst Electr Electron Eng) Trans Biomed Eng 2006;53:1346—56. https://doi.org/10.1109/TBME.2006.875664. http://ieeexplore.ieee.org/document/1643403/.

[59] Ainsworth BE, Haskell WL, Whitt MC, Irwin ML, Swartz AM, Strath SJ, O'Brien WL, Bassett DR, Schmitz KH, Emplaincourt PO, Jacobs DR, Leon AS. Compendium of physical activities: an update of activity codes and MET intensities. Med Sci Sports Exerc 2000;32:S498—504. http://www.ncbi.nlm.nih.gov/pubmed/10993420.

[60] Rowlands AV, Yates T, Olds TS, Davies M, Khunti K, Edwardson CL. Sedentary sphere, medicine & science in sports & exercise, vol. 48; 2016. p. 748—54. https://doi.org/10.1249/MSS.0000000000000813. http://www.ncbi.nlm.nih.gov/pubmed/26559451Ωhttps://insights.ovid.com/crossref?an=00005768-201604000-00022.

[61] Perez-Pozuelo I, White T, Westgate K, Wijndaele K, Wareham NJ, Brage S. Diurnal profiles of physical activity and postures derived from wrist-worn accelerometry in UK adults. J Measure Phys Behaviour 2019:1—11. https://doi.org/10.1123/jmpb.2019-0024.

[62] Dunstan DW, Howard B, Healy GN, Owen N. Too much sitting — a health hazard, diabetes research and clinical practice 2012;vol. 97:368—76. https://doi.org/10.1016/j.diabres.2012.05.020. http://www.ncbi.nlm.nih.gov/pubmed/22682948Ωhttp://linkinghub.elsevier.com/retrieve/pii/S0168822712002082.

[63] Lambrecht S, Del-Ama AJ. Human movement analysis with inertial sensorsvol. 4. Springer International Publishing; 2014. p. 305—28. https://doi.org/10.1007/978-3-642-38556-8_16.

[64] A. Bulling, 33 A Tutorial on Human Activity Recognition Using Body-Worn Inertial Sensors (????). https://doi.org/10.1145/2499621. Crossref Partial Au1 ATL.

[65] Plotz T, Guan Y. Deep learning for human activity recognition in mobile computing. Computer 2018;51:50—9. https://doi.org/10.1109/MC.2018.2381112. https://ieeexplore.ieee.org/document/8364643/.

[66] Radu V, Tong C, Bhattacharya S, Lane ND, Mascolo C, Marina MK, Kawsar F. Multimodal deep learning for activity and context recognition. Proc ACM Interact, Mob, Wearable Ubiquitous Technol 2018;1:157.

[67] Bulling A, Ward JA, Gellersen H. Multimodal recognition of reading activity in transit using body-worn sensors. Trans Appl Percept 2012;9:2.

[68] Hemminki S, Nurmi P, Tarkoma S. Accelerometer-based transportation mode detection on smartphones. In: Proceedings of the 11th ACM Conference on embedded networked sensor systems, ACM; 2013. p. 13.

[69] Guo H, Chen L, Peng L, Chen G. Wearable sensor based multimodal human activity recognition exploiting the diversity of classifier ensemble. In: Proceedings of the 2016 ACM International joint Conference on pervasive and ubiquitous computing, ACM; 2016. p. 1112—23.

CHAPTER 4

Artificial intelligence/machine learning solutions for mobile and wearable devices

Zhenxing Xu, PhD [1], Bin Yu, PhD [2], Fei Wang, PhD [1]

[1]Department of Population Health Sciences, Weill Cornell Medicine, Cornell University, New York, NY, United States; [2]American Air Liquide Inc New York, NY, United States

Contents

The development of mobile and wearable devices such as smartphones provides an opportunity to collect high-density physiological, biological, or behavioral data about people's daily life. Coupled with the progress of artificial intelligence (AI) and machine learning (ML), the massive amount of data collected from these devices can help inform impending conditions immediately, such as identifying data trends over time and spotting any suspecting deviations. These technical revolutions bring new promises to healthcare with the hope of improving people's quality of life. In this chapter, we summarize popular mobile and wearable devices, popular data types they captured, and the associated AI/ML techniques for analyzing them.

1. Mobile and wearable devices

Mobile and wearable devices are receiving substantial attentions in recent years because of their potential of capturing continuous, real-time physiological signals such as heart rate, blood pressure, skin temperature,

Digital Health
ISBN 978-0-12-820077-3
https://doi.org/10.1016/B978-0-12-820077-3.00004-3

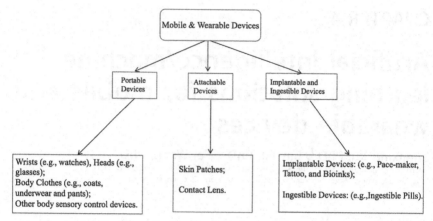

Figure 4.1 The hierarchy classification of mobile and wearable devices.

respiratory rate, and body motion, which can be used to extract clinically relevant information for determining health status [1,2]. With the rapid development of electronic, biocompatible materials, and nanomaterials, mobile and wearable devices have gradually been developed in diverse forms such as accessories, integrated clothing, and body insertions [3]. Generally, mobile and wearable devices can be divided into three categories [4]: (1) portable devices; (2) attachable devices; and (3) implantable and ingestible devices, which are demonstrated in Fig. 4.1. The portable devices mainly contain popular portable medical and healthcare devices worn on body parts such as wrists (e.g., watches), heads (e.g., glasses), body clothes (e.g., coats, underwear, and pants), and other body sensory control devices [5,6]. Attachable devices are mainly in the forms of wearable skin patches [7] with skinlike adaptability and flexibility, which are seen as the next-generation personal portable healthcare devices and have the potential to obtain more accurate and reliable sensing information without being disturbed by movement. Researchers have used wearable patches attached on human skin to obtain the signals about cardiovascular, sweat, and temperature. Recently, a wearable contact lens optical sensor has been developed to monitor glucose in physiological conditions [7]. The implantable and ingestible devices mainly include implantable devices and ingestible pills. The implantable devices have recently been used to perform wireless medical measurements for diagnosing and treating diseases by detecting changes in the body such as rejection of transplantation [4]. Representative implantable devices include pacemaker [8], tattoo [9], and bioinks [10]. Pacemaker can be used to treat heart arrhythmias by bringing the rhythms back to normal using low-energy electrical pulses when irregular heartbeats are discovered. It is the most popular implanted medical device for patients with heart diseases [8]. Tattoos are used for monitoring emotions and vital signs, which are adapted to various skin textures and provide natural wearability for the patient and precise data to the physician [9]. Bioinks (biosensitive inks) adopt a simple chemical reaction to obtain the changes in terms of certain elements (e.g., sodium) in the body, which does not require power to process and transmit data. The ingestible pills are mainly suitable for endoscopy because they pass through the digestive system. Currently, the ingestible pills are mainly used to monitor the precise time in terms of the taken drug. When the ingestible pills reach the stomach, a chemical reaction with the stomach fluid will happen and then a signal is sent to the patch worn on the body [2].

2. Different types of data collected by mobile and wearable devices and applications

The mobile and wearable devices can collect different types of data, which are summarized as follows:

(1) Physiological data collected by wearable devices: Psychological data usually include heart rate, blood glucose, blood pressure, respiration rate, body temperature, blood volume, sound pressure, photoplethysmography, electroencephalogram, electrocardiogram, blood oxygen saturation, and skin conductance [2,11,12]. These physiological data can be used for multiple applications. More details of these applications and studies are shown in Table 4.1.

  (a) Preventive health: The user takes a wearable like a wristwatch clamped on his/her wrist to obtain physiological data such as heart rate and blood pressure, which are fed to the platform with an AI engine that uses ML methods to detect abnormalities if any in the wearer's body. The wearable provides a report for users in terms of potential conditions such as blocked arteries that could lead to heart stroke or cancerous tumor at a very early stage if the abnormalities are detected. Preventive disease wearables make more contributions to the healthcare industry and reduce our annual healthcare budget cost [11].

  (b) Medical consultation: On detection of an abnormality, the use can provide the report to their consulting physician or an AI doctor. AI doctor is generally a standalone platform with an AI engine that

Table 4.1 Summary of physiological data and applications.

Common physiological data	Device	Applications	An example of related study
Heart rate	Pulse oximeter/ skin electrodes	Preventive health and monitoring stress	[11]
Blood glucose	Skin patches	Monitoring chronic disease conditions	[12]
Blood pressure	Wrist-mounted devices	Preventive health	[17]
Respiration rate	Body clothes	Preventive health	[18]
Body temperature	Skin patches	Preventive health and monitoring stress	[19]
Blood volume	Skin patches	Other applications (e.g., emotion detection)	[16]
Photoplethysmography	Wrist-mounted devices	Preventive health and monitoring stress	[20]
Electroencephalogram	Scalp-placed electrodes	Preventive health and medical consultation	[14]
Blood oxygen saturation,	Skin patches	Medical consultation and medication management	[21]
Electrocardiogram	Skin/chest electrodes	Preventive health and medical consultation	[13]
Other types (monitoring medication)	Ingestible devices	Medication management	[15]

collects a great amount of medical knowledge from multiple medical databases, which usually can detect ailments faster than an actual doctor. An AI doctor empowers users by providing them the medical knowledge to become more aware about certain diseases and personal health. In addition, doctors can save time by reading the reports produced by wearable devices and focus on finding solutions to their patients' diseases [13,14].

(c) Medication management: Wearable devices usually collect physiological data, produce health reports, and provide prescriptions to users if there are abnormalities. The users can choose some pills based on the prescriptions. In addition, ingestible devices (e.g., ingestible pills) can monitor the precise time at which any drug is taken [15]. The device can be integrated with other wearable health apps and has the potential to remind users when it is time to take a medicine.

(d) Monitoring chronic disease conditions: Wearable devices can be used to monitor minute-to-minute chronic conditions of the patients, which is very hard for the nurses and the doctors [12]. For instance, for a diabetic patient, it is dangerous to have large fluctuations in its blood sugar level. Wearable devices are used to track these fluctuations, improve the insulin treatment to be tailored with high precision, optimize the diabetic control, and prevent dangerous extreme levels of blood sugar. For patients who suffer from palpitations, the wearable devices placed on a patient's wrist can detect small and infrequent disturbances, which is almost impossible for the nurses and the doctors to discover.

(e) Monitoring stress: The heart rate variability information can be collected by wearable devices, which can be used to provide insight into people's mental well-being and body stress. For example, emotions such as "Sad," "Dislike," "Joy," "Stress," "Normal," "No-Idea," "Positive," and "Negative" are recognized based on physiological data, and provide evaluation to the users' mental status [16]. Some breathing exercises that lean on onboard heart rate monitors can also obtain benefits from physiological data such as respiration rate, which provide insights into letting users take control of how they handle stress and help users to obtain a calmer state.

(2) Sports data collected by wearable devices: Wearable devices like smartphones, watches, and wristbands can collect sport-related data such as acceleration information, rotation speed, and running steps. These sport-related data are used to monitor the motion state, recognize sport behavior, and monitor sports. More details in terms of these applications and studies are shown in Table 4.2.

(a) Monitoring motion state: The acceleration data collected by wearable devices can be used for posture and movement recognition and fall detection and balance control evaluation, providing warnings for special cases, and monitoring the elder [22], which offers timely treatment for users after falling, improves the quality of life of some groups such as the elder, and achieves healthy aging.

Table 4.2 Summary of sports data, environmental data, and their applications.

Common sport data or environmental data	Device	Applications	An example of related study
Acceleration information	Smartphones, watches, and wristbands	Monitoring motion state	[22]
Running step and acceleration information	Wristbands	Recognizing sport behavior and monitoring motion state	[23]
The level of sweat	Body clothes	Monitoring sport	[24]
Rotation speed and acceleration information	Wrist-mounted devices and other body sensory devices such as head	Recognizing sport behavior and monitoring sports	[28]
Magnetic field strength and GPS	Smartphones	Localizing position and tracking	[29]
GPS and environmental luminous flux	Head devices such as glasses	Improving security	[25]
GPS	Body clothes	Localizing position and tracking	[27]

(b) Recognizing sport behavior: The recognition of sport behavior based on wearable devices usually considers the cycling, running, climbing stairs, and other specific types of movements, which can be used for the comprehensive monitoring of athletes' behavior in activities, family behavior analysis, security monitoring, private data processing, and preventing injuries [23]. For example, for patients with Parkinson's disease, human activity recognition is useful for collecting daily activity logs and detecting gait status to evaluate the patient's condition and progress in recovery.

(c) Monitoring sports: The aim of the sport monitoring mainly focuses on avoiding excessive exercise and even heart attacks by considering sport data such as rotation speed, which provides guilds for users by detecting whether the user has reached an appropriate exercise plan and it is proper for long-term physical exercise. For example, by detecting the level of sweat using intelligent socks, users can obtain guilds during exercising [24].

(3) Environment data collected by wearable devices: The wearable devices such as watches and glasses can collect environmental data such as global positioning system (GPS) coordinates, magnetic field strength, and environmental luminous flux, which can be provided for specific applications. More details in terms of these applications and studies are shown in Table 4.2.

(a) Improving security: The wearable LED lights are designed by Lumenus [25] company for runners, bicyclists, and motorcyclists, which consider multiple sources such as GPS to control the color, brightness, and animation of wearable LED lights. The system can provide guilds according to the direction of where a bicyclist is heading and automatically offer a turn signal, alerting the drivers around the user.

(b) Localizing position and tracking: The wearable tracking devices with GPS can offer benefits for specific groups such as children, which keep track of young children and avoid missing. In addition, compared to providing phones for children, some wearable tracking devices such as intelligent underwear and pants are safer and more convenient [26,27].

3. The components of mobile- and wearable devices—based applications

In the last decade, mobile and wearable devices have generated substantial impacts on different aspects of our daily lives such as preventive health, medical consultation, medication management, etc. Most of these application systems contain three parts [30]: sensor module, wireless transmission module, and back-end monitoring module. The general workflow for mobile and wearable devices is shown in Fig. 4.2.

The sensor module is used to measure and collect physiological information, such as human body temperature and heart rates, from the user's body. Some basic analytical processes (e.g., signal filtering) will be performed on the collected data. Wireless transmission module is used to enable the transmission of the information collected via the sensor module from the mobile and wearable devices to a terminal device. The wireless

Figure 4.2 The general workflow for mobile and wearable devices.

transmission modules usually have large-distance, low-power, and low-cost design. There are several popular types of wireless transmission devices: bluetooth, infrared, radio-frequency identification, and near-field communication technology [31]. For the back-end monitoring module, the primary functions are to record the user's physiological information, analyze the collected information, and set the parameters of wireless transmission devices. It will generate alerts to specific users by a control buzzer based on the evaluated results in terms of the risk level of certain diseases (e.g., heat stroke) [30]. Among these modules, the processing of the collected data usually involves AI and ML algorithms, which make the mobile and wearable devices smarter and more intelligent. For example, doctors can utilize virtual reality glasses to watch X-rays without leaving the operating room and getting distracted from the operation process. Clinicians can obtain patient health reports produced by ML algorithms based on patients' physiological information to get a comprehensive understanding of the patient health conditions.

4. The ML solutions for mobile and wearable devices

Mobile and wearable devices can collect diverse physiological signals such as heart rate (HR), body temperature (BT), respiration rate (RESP), blood pressure (BP), electroencephalogram (ECG), electrocardiogram (EEG), photoplethysmography (PPG), Galvanic skin response (GSR), oxygen saturation (SpO2), blood glucose (BG), and body mass index (BMI) [11,12]. These collected data are usually referred to as streaming data. Compared with the data collected in clinic settings such as electronic health records (EHRs) [32], the volume of these data is much bigger and they contain more noises due to irregular body movements. There are two common methods used to deal with these data [33]: signal processing techniques and ML-based techniques. Signal processing techniques can derive effective representations from both time and frequency domains. For example, in artifact processing, frequency-specific filters (e.g., low-pass, high-pass, band-pass, band-stop) are used to remove muscle artifact [34]. For the sleep study with EEG signals, Kalman filters can be used to predict future values of the time series data and detect significant deviations [35]. ML-based methods usually construct a model based on the training data for the classification or regression problem [36,37]. Signal processing techniques are usually integrated with ML techniques [38] to improve the model performance. The popular AI/ML techniques are summarized in Fig. 4.3.

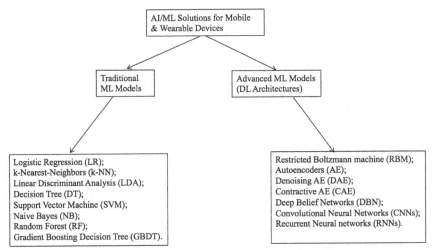

Figure 4.3 Multiple popular AI/ML solutions for mobile and wearable devices.

Conventional ML-based studies typically contain three main steps for analyzing streaming data collected from mobile and wearable devices (1) segmentation: using partition techniques (e.g., sliding window approach) to split raw streaming data into discrete (could be overlapping) segments or pieces for further processing; (2) feature extraction: applying feature engineering methods to derive effective feature representations based on raw input data. The extracted features are usually domain dependent, such as time-domain (e.g., signal amplitude, zero-crossings, mean, standard deviation, higher moments), frequency-domain (e.g., energy bins, Fourier coefficients, frequency peaks, power spectral density), or time-frequency domains (e.g., wavelet coefficients). (3) Decision: based on the feature vectors derived from the previous two steps, an ML model will be trained to produce a decision for testing data. In the following we describe several representative conventional ML techniques. More details of traditional ML methods, wearable device, data, and their applications are summarized in Table 4.3.

(1) k-Nearest-Neighbors (k-NN): k-NN is a nonparametric ML approach for classification and regression [39]. The decisions of testing data are usually produced by observing the k closest training data according to the Euclidean distance evaluated in the feature space, this algorithm can be viewed as a lazy learner. The selection of value k is sensitive to the data distributions. A larger k would decrease the sensitivity because of the increase of bias and noise data, where the smaller value k would

Table 4.3 Summary of traditional ML methods, wearable device, data, and their applications.

Methods	Devices	Data	Applications	Selected related study
k-NN	Wrist wearable device.	Acceleration; velocity; Displacement; Skin temperature.	Fall detection; Detection of opioid use.	[33,40,41,56]
LR	Smartphones; Wearable chest belts; Scalp-placed electrodes.	Heart rate; EEG signals.	Stress level assessments; Seizure detection.	[43,44]
DT	Wearable shoe sensor; smart clothing.	ECG signals; respiration rate; acceleration; Temperature; Audio; Blood glucose.	Action and gesture recognition; diabetes diagnosis; Assessment of health status of patients with chronic obstructive pulmonary disease.	[46–48]
SVM	Scalp-placed electrodes; Wearable chest belts.	EEG signals; ECG signals.	Seizure detection; arrhythmia detection.	[50,51]
RF	Wearable gadget and smartphone.	Heart rate; temperature; blood pressure; Acceleration; velocity; Displacement.	Heart disease condition monitoring; Fall detection.	[53,54]
GBDT	Wrist wearable device.	Electrodermal activity, skin temperature, and tri-axis acceleration data.	Detection of opioid use.	[56]

result in an unstable decision by increasing variance. The main disadvantage of the k-NN algorithm is the complexity in searching the nearest neighbors for each testing data. Several studies about fall detection based on wearable data collected via a wrist wearable device use k-NN as a classifier to detect the status of the movement [33,40,41].

(2) Logistic Regression (LR): LR classifier is a linear classification method [42] aiming at learning the optimal set of coefficients to describe the relationship between the label and a set of feature variables. The main advantage of the LR classifier is relatively robust, flexible, and easy to interpret. However, the performance it can achieve is usually limited due to the simple linear assumption. LR models have been used in many applications involving mobile and wearable devices such as stress level assessments with information collected from smartphones [43] and seizure detection from EEG signals [44].

(3) Decision Tree (DT): The goal of DT is to construct a set of rules for iterative partitioning the feature spaces to achieve the final decision [45]. Each node in a DT corresponds to a sample subset and the branch encodes a specific feature partitioning rule to obtain the sample subset in the node. There are several algorithms to produce optimal trees such as ID5 and CART (Classification And Regression Tree). There are many advantages of DT, such as its robustness with respect to the feature types and easy to interpret. However, DT classifier is not very stable, which causes a large change in the structure of the optimal decision tree when there is a small change in the data. Moreover, information gain in DT would be biased if the variables of data contain categorical variables. DT has been applied in scenarios of using the data collected by wearable sensors for action and gesture recognition from wearable sensors [46], diabetes diagnosis [47], assessment of health status of patients with chronic obstructive pulmonary disease [48], etc.

(4) Support Vector Machine (SVM): SVM is a linear model designed for separating the data from two classes with the maximum margin, which is defined as the samples from different classes that are closest to the decision function (a.k.a. support vectors) [49]. To further enhance the power of SVM, kernel trick has been introduced to inject nonlinearities to SVM. SVM has been a popular model and demonstrated to be able to achieve state-of-the-art performance in many applications before the deep learning (DL) era. It has also been applied in many scenarios in mobile health [50] and wearable devices [51].

(5) Random Forest (RF): RF is a model integrating multiple DTs, with each trained on a randomly created subset of training data. The final decision of the RF can be obtained by aggregating (typically by majority voting) the decision from each individual tree [52]. During the process of building RF, bootstrap sampling and feature splitting are needed to create the random subsets for learning the base DTs. Through the ensemble strategy, RF effectively reduces the performance variance DT and greatly improves the algorithm stability. RF is still one of the most popular ML models nowadays because of its superior performance and interpretability. RF has been widely used in mobile- and wearable device—related applications such as heart disease condition monitoring [53] and fall detection [54], etc.

(6) Gradient Boosting Decision Tree (GBDT): GBDT is another ensemble DT method based on gradient lifting algorithm [55]. It iteratively builds an ensemble of weak decision tree learners through boosting, which is able to upgrade weak learners to strong learners. Compared with RF, GDBT can typically achieve the same level performance with less number of trees, which thus is more efficient. GBDT has achieved state-of-the-art performance on many mobile- and wearable device—related applications. Mahmud et al. [56] compared the performance of three ML methods including k-NN, DT, and GBDT in terms of the detection of opioid intake based on multiple types of data such as electrodermal activity, skin temperature, and tri-axis acceleration data acquired from wrist wearable devices, and GBDT showed the best performance.

One important aspect for the success of those conventional ML techniques is effective feature engineering, which is usually expensive, time-consuming, and labor-intensive. DL is the new generation of ML methods that adopts an end-to-end learning scheme to automatically identify complex patterns from data [57]. DL techniques have achieved state-of-the-art performance in many domains including computer vision, speech recognition, natural language processing, computational biology and biomedicine, etc. There have also been an increasing number of studies applying DL models for analyzing physiological signals collected from mobile and wearable devices. In the following we will review them according to the model types. More details in terms of DL methods, wearable device, data, and their applications are summarized in Table 4.4.

(1) Restricted Boltzmann machine (RBM): RBM is a DL architecture learning effective representations for input data in an unsupervised way [58]. An RBM is a generative stochastic model, which learns a

Table 4.4 Summary of DL methods, wearable device, data, and their applications.

Methods	Devices	Data	Applications	Selected related study
RBM	Smartphone.	PPG signals.	Identification of PPG signals.	[61]
AE	Wearable chest or leg belts.	Acceleration; velocity; Displacement.	Identification of unseen falls; Locomotion activity recognition.	[63,64]
CNNs	Scalp-placed electrodes; Smartphone.	Skin images; EEG signals; Voice signals.	Melanoma classification; Seizure prediction; Parkinson's disease identification.	[36,37,66,67]
RNNs	Wrist wearable device; Smartphone.	Daily step counts, sleep duration, and interactions information with specific apps.	Predicting traits associated with wellness (BMI, weekly alcohol intake), lifestyle (food diary habits and scale utilization); Investigating behavioral traits (activity on a fitness-related social network).	[69]

probability distribution over the input space and generates latent representations. An RBM is a variant of Boltzmann machines (BM) [59] and can be seen as a stochastic neural network, which consists of one layer of observed units, one layer of hidden units plus a bias unit. All units in BM are fully connected, while there are no connections in RBM between any two observed units or any two hidden units, the links only exist between observed and hidden nodes. Deep Belief Networks (DBN), which is one of the RBM variants, has been successfully applied in EHRs embedding and portable neurophysiological signals analysis [60]. Jindal et al. [61] proposed a two-stage technique by integrating RBM and DBN to identify PPG signals in clinical, e-health, and fitness environments. The proposed model was tested on the TROIKA dataset that was collected by wearable biosensors and obtained an accuracy of 96.1%.

(2) Autoencoders (AE): AE is a special variant of the feedforward neural network (FNN) [62], which stacks the neurons layer by layer in a fully connected feedforward manner. The goal of AE is to learn new (usually lower-dimensional) representations of input data in an unsupervised manner. An AE contains two parts: encoder and decoder. Encoder compresses the input data into a latent representation vector. Decoder reconstructs the original data samples from the latent representations learned by the encoder. The parameters in the AE architecture are obtained by minimizing the reconstruction loss. A deep AE structure with more hidden layers has the potential to increase the level of abstraction of learned features. There are multiple variants based on AE structure such as stacked AE (SAE), denoizing AE (DAE), variational AE (VAE), contractive AE (CAE), etc. In order to improve the performance of fall detection, S. Khan et al. [63,64] proposed an ensemble AE model to learn generic features from different channels of wearable sensor data. The proposed model obtained better results on two activity recognition datasets compared to traditional AE models and other two standard one-class classification techniques. Gu et al. [63,64] used the AE model to identify locomotion activity based on sensor data extracted from four sensors including accelerometer, gyroscope, magnetometer, and barometer, and achieved a higher accuracy.

(3) Convolutional Neural Networks (CNNs): CNNs are a specific category of neural networks originally designed for image analysis inspired by the mechanism of human visual pathway [65]. A CNN architecture typically consists of three building blocks: convolution—activation layer, pooling layer, and fully connected layer. The convolution—activation layer applies multiple local convolution filters over the original image to obtain the feature maps which will go through nonlinear activation functions. The main purpose for pooling layers is to perform downsampling to the feature maps to improve the robustness and generalization ability of the CNN. The fully connected layer is used to perform final prediction or classification (e.g., through softmax function). Different from FNN where the nodes in each layer is fully connected to the previous layer, the convolution filters in CNN focus on local translational invariant feature patterns, which mimics the local

receipt fields of visual axons and makes the model computational tractable. CNNs have been widely used in applications including melanoma classification with skin images [66], seizure prediction with EEG signal analysis [36,37], Parkinson's disease identification from voice signals [67], etc.

(4) Recurrent Neural networks (RNNs): RNNs are specifically designed for processing sequential data such as natural language and speech [68]. The fundamental module in the RNN is a recurrent cell which takes the output from the last timestamp and the new input at current timestamp to produce the current output. There are two popular variants of RNN: long–short term memory and gated recurrent unit. These two models are designed to alleviate vanishing of gradients and learn long-term temporal dependencies within time-series data or streaming data. The primary difference between these two models lies in the different ways to map the input to the output through recurrent neurons. As the streaming data captured from mobile and wearable devices are sequential in nature, Quisel et al. proposed an RNN framework to analyze millions of mHealth data streams [69].

Although DL techniques have demonstrated success in multiple applications and become more and more popular, there is no free lunch. It is necessary to understand differences between traditional ML and DL techniques when facing real-world problems. For example, since there is a large number of parameters that need to be trained in DL techniques, the better performance can be usually obtained as the scale of data increases. On the other hand, if there is not enough training data, traditional ML techniques with their handcrafted rules prevail in this scenario. However, in many scenarios acquiring large amounts of training data is expensive and sometimes even impossible, especially in healthcare and medical settings [70]. Also features used in traditional ML techniques are usually extracted based on domain knowledge from experts, which play an important role in the performance of models. The process of extracting features are usually challenging and expensive in terms of time and professional knowledge. DL approaches work in an end-to-end way and do not require any human intervention. This usually makes the features extracted with DL approaches difficult to explain and interpret [71].

5. Case study: the application of mobile and wearable device data in the COVID-19 pandemic

The novel coronavirus disease 2019 (COVID-19) has caused a large number of infected people and deaths, produced serious impacts on people's life, and resulted in considerable economic loss [72]. Nonpharmaceutical interventions, such as large-scale testing and contact tracing, restricting international and domestic travels, social distancing, and locking down specific regions, play an important role in controlling the COVID-19 pandemic. Mobile phone data have proven to be a valuable data source for capturing human mobility and social interactions [72]. In this section, the value and contribution of mobile phone data in terms of controlling the COVID-19 pandemic are summarized briefly. In particular, three different aspects in terms of applications of mobile phone data for the COVID-19 pandemic are introduced, including contact tracing, detecting and screening, and monitoring behavioral changes.

(1) **Contact tracing**: The application of contact tracing on the basis of mobile phone data can be divided into three stages [72], including the early recognition and initiation phase of the COVID-19 pandemic, the acceleration phase, and the deceleration and preparation phase. For the first stage, mobile phone data are used to discover the locations and social networks of infected individuals, which are further employed for situational analysis and perform the fast detection for the contacts. In this stage, origin—destination matrices models are especially used to evaluate the COVID-19 pandemic situation, which mainly focus on assessing the mobility of the population. For the second stage, aggregated mobile phone data are used to evaluate the efficacy of implemented measures such as mobility restrictions. In addition, a great amount of mobility information is used to build epidemiological models such as the estimations of dwell and hotspots, which make contributions to the explanation of the spread of the COVID-19 pandemic and optimize resource allocation such as respirators. For the third stage, after the COVID-19 pandemic has subsided, post hoc analysis of the influence of different measures for controlling the disease is performed. In this stage, multiple digital contact-tracing apps are developed and deployed, such as the Singaporean smartphone app TraceTogether, and the Korean smartphone app Corona 100m. Recently, Apple and Google have worked together to develop a system [72], which supports Bluetooth-based privacy-preserving proximity tracing across iOS and Android smartphones. Although contact-tracing apps have made important contributions to addressing the COVID-19 pandemic by tracing contact, the use of these apps remains relatively scarce, which might involve the issue of privacy, data protection, and civil liberties.

(2) **Detecting and screening**: Due to high transmissibility and no effective vaccine or therapy, it becomes very important to widely detect and screen COVID-19 in subjects, especially for an asymptomatic individual with COVID-19 exposure. However, the limitations of medical diagnosis materials and complex processes such as using nasopharyngeal sampling for COVID-19 reverse transcription polymerase chain reaction (RT-PCR) pose a challenge to the detection for COVID-19 in all subjects [73]. Wearable biosensors make it possible by continuously monitoring multidimensional physiological data for early detection of COVID-19 clinical progression. In particular, wearable devices collect skin temperature, respiratory rate, blood pressure, pulse rate, blood oxygen saturation, and daily activities, which are automatically transferred in real time to AI platform such as BiovitalsAnalyticsEngine [73] by smartphone application and perform analysis, and report the result to the hospital and government department. They can take measures such as isolation and quarantine to control the spread and outbreak of COVID-19 based on these results. In addition, some governments such as China use health code for individual's travel, which is developed on a large amount of contact history and biometric data collected by wearable devices such as smartphones. Currently, the application of China's health tracking QR codes has been shown to considerably reduce the new infected subjects and play an important role in the country's successful containment of the COVID-19 pandemic, which provides some insights into controlling disease for other countries to consider and follow.

(3) **Monitoring behavioral changes**: Nonpharmaceutical interventions such as keeping social distance and full lockdown have proven to be critical for considerably reducing the new confirmed case [72]. However, due to the different cultures in different countries or regions, these interventions might produce different responses to the population such as changing their behavioral patterns. In addition, longer periods of isolation pose a risk to the mental health changes such as resulting in more anxious and depressed [74]. It is urgently required for researchers and government departments to investigate and address the response of these interventions. The increasing availability of wearable devices plays an important role due to the ability of efficiently collecting near-real-time high-resolution data from large numbers of individuals and greatly facilitating remote monitoring of behavior. For example, a recently developed open-source mobile health platform

RADAR-base is used to collect individual data from smartphone and wearable devices and provide these data for researchers [75]. These individual data usually include time staying at home, the distance traveled from home, the number of Bluetooth-enabled nearby devices (as a proxy for physical distancing), step counts, heart rate, sleep duration, bedtime, phone unlock duration, and social app use duration [75]. Multiple interesting findings are obtained such as spending more time on social media apps, sleeping more, and having lower heart rate. In addition, some researchers have investigated mental changes in specific groups such as college students and found that individuals were more sedentary, anxious, and depressed during the COVID-19 pandemic [74].

6. Conclusion

Mobile and wearable devices are becoming more and more important in healthcare. AI/ML techniques have been successfully applied in analyzing the data captured from mobile and wearable devices and triggered many applications. In this chapter, we have reviewed the popular mobile and wearable devices along with the common AI/ML techniques. However, there are still many technical challenges and future directions for utilization and deployment of mobile and wearable devices. For example, (1) inaccurate sensor data measurement: mobile and wearable devices should not only be able to collect users' physiological data accurately in the lab but also should achieve a high accuracy in wild environmental conditions such as wet and warm. (2) Individual differences: mobile and wearable devices can provide health information based on users' physiological data. There is a large variation on such data from patient to patient. Therefore, effective exploration of individual differences to improve precision care and medicine remains to be investigated. (3) Limited energy supply and storage: AI techniques usually need massive calculation and storage resources when analyzing users' physiological data. (4) Language translation function: the new generation wearable devices with language translation function has the potential to provide more benefits for humans all over the world. The wearable medical devices with language translation function can assist doctors in discussing treatment options with patients who might need a translator. In addition, medical emergencies might occur at any time and any place. The wearable devices might help users chat with strangers on the street in case of a medical emergency, where the users do not speak the same language as those around the users. (5) Hearables: wearable hearing aids devices provide many benefits for humans such as hearing impaired. However, existing hearables are prohibitively expensive. In addition, current devices are not able to automatically dampen the noise in a muddled conversation environment such as in a busy restaurant. The hearables devices can also be integrated with haptic feedback to remind users to take pills or check blood pressure. (6) Miniaturized wearable devices: the designers should consider lighter-weight wearable designs for next-generation wearables by reducing the size of current version and using an integrated manner to combine multiple sensors, and at the same time keeping their efficacy in a small area. In addition, in order to obtain reliable measurements, many wearable devices may require to be placed on very specific body placement or body postures.

Acknowledgment

We would like to acknowledge the support from American Air Liquide Inc. for this work.

References

[1] Kim J, Campbell AS, de Ávila BE-F, Wang J. Wearable biosensors for healthcare monitoring. Nat Biotechnol 2019;37:389–406.
[2] Sim I. Mobile devices and health. N Engl J Med 2019;381:956–68.
[3] Hwang I, Kim HN, Seong M, Lee S-H, Kang M, Yi H, et al. Multifunctional smart skin adhesive patches for advanced health care. Adv Healthc Mater 2018;7:e1800275.
[4] Guk K, Han G, Lim J, Jeong K, Kang T, Lim E-K, et al. Evolution of wearable devices with real-time disease monitoring for personalized healthcare. Nanomaterials 2019;9. https://doi.org/10.3390/nano9060813.
[5] Han G. Latest progresses in developing wearable monitoring and therapy systems for managing chronic diseases, vol. 1802; 2018. p. 01747. arXiv preprint arXiv.
[6] Gonzales WV, Mobashsher A, Abbosh A. The progress of glucose monitoring—a review of invasive to minimally and non-invasive techniques, devices and sensors. Sensors 2019:800. https://doi.org/10.3390/s19040800.
[7] Elsherif M, Hassan MU, Yetisen AK, Butt H. Wearable contact lens biosensors for continuous glucose monitoring using smartphones. ACS Nano 2018;12:5452–62.
[8] Waln O, Jimenez-Shahed J. Rechargeable deep brain stimulation implantable pulse generators in movement disorders: patient satisfaction and conversion parameters. Neuromodulation 2014;17:425–30. discussion 430.
[9] Kim J, Jeerapan I, Imani S, Cho TN, Bandodkar A, Cinti S, et al. Noninvasive alcohol monitoring using a wearable tattoo-based iontophoretic-biosensing system. ACS Sens 2016:1011–9. https://doi.org/10.1021/acssensors.6b00356.
[10] Liu X, Yuk H, Lin S, Parada GA, Tang T-C, Tham E, et al. 3D printing of living responsive materials and devices. Adv Mater 2018;30. https://doi.org/10.1002/adma.201704821.
[11] Baig MM, Gholamhosseini H. Smart health monitoring systems: an overview of design and modeling. J Med Syst 2013;37:9898.

[12] Lee H, Choi TK, Lee YB, Cho HR, Ghaffari R, Wang L, et al. A graphene-based electrochemical device with thermoresponsive microneedles for diabetes monitoring and therapy. Nat Nanotechnol 2016;11:566–72.

[13] Luz EJ da S, Schwartz WR, Cámara-Chávez G, Menotti D. ECG-based heartbeat classification for arrhythmia detection: a survey. Comput Methods Progr Biomed 2016;127:144–64.

[14] Craik A, He Y, Contreras-Vidal JL. Deep learning for electroencephalogram (EEG) classification tasks: a review. J Neural Eng 2019;16:031001.

[15] Chai PR, Rosen RK, Boyer EW. Ingestible biosensors for real-time medical adherence monitoring: MyTMed. Proc Annu Hawaii Int Conf Syst Sci 2016;2016:3416–23.

[16] Khan AM, Lawo M. Recognizing emotion from blood volume pulse and skin conductance sensor using machine learning algorithms. In: XIV mediterranean conference on medical and biological engineering and computing, vol. 2016; 2016. p. 1297–303. https://doi.org/10.1007/978-3-319-32703-7_248.

[17] Hsu Y, Young DJ. Skin-surface-coupled personal health monitoring system. Sensors 2013;2013:1–4. IEEE.

[18] Wu D, Wang L, Zhang Y-T, Huang B-Y, Wang B, Lin S-J, et al. A wearable respiration monitoring system based on digital respiratory inductive plethysmography. Conf Proc IEEE Eng Med Biol Soc 2009;2009:4844–7.

[19] Yoon S, Sim JK, Cho Y-H. A flexible and wearable human stress monitoring patch. Sci Rep 2016;6:23468.

[20] Hwang S, Lee S. Wristband-type wearable health devices to measure construction workers' physical demands. Autom ConStruct 2017;83:330–40.

[21] Pang Z, Tian J, Chen Q. Intelligent packaging and intelligent medicine box for medication management towards the Internet-of-Things. In: 16th international conference on advanced communication technology; 2014. https://doi.org/10.1109/icact.2014.6779193.

[22] Yang C-C, Hsu Y-L. A review of accelerometry-based wearable motion detectors for physical activity monitoring. Sensors 2010;10:7772–88.

[23] Santos-Gago JM, Ramos-Merino M, Vallarades-Rodriguez S, Álvarez-Sabucedo LM, Fernández-Iglesias MJ, García-Soidán JL. Innovative use of wrist-worn wearable devices in the sports domain: a systematic review. Electronics 2019;8:1257.

[24] Zhu M, Shi Q, He T, Yi Z, Ma Y, Yang B, et al. Self-powered and self-functional cotton sock using piezoelectric and triboelectric hybrid mechanism for healthcare and sports monitoring. ACS Nano 2019;13:1940–52.

[25] ioLIGHT by Lumenus: A smart bike light. [cited 27 Aug 2020]. Available: https://www.lumenus.com/.

[26] Ahmed N, Zade PL. Circularly polarized wearable textile antenna with defected ground structure for GPS application. In: 2017 international conference on innovations in information, embedded and communication systems (ICIIECS); 2017. p. 1–4.

[27] Banchongraksa B, Truong J, Chieh LC, Yacoub M, Meteekotchadet P, Daim T. Personal transformation: wearable GPS device for children. Dig Transform 2019:299–327. WORLD SCIENTIFIC.

[28] Aroganam G, Manivannan N, Harrison D. Review on wearable technology sensors used in consumer sport applications. Sensors 2019;19. https://doi.org/10.3390/s19091983.

[29] Ashraf I, Hur S, Park Y. MagIO: magnetic field strength based indoor- outdoor detection with a commercial smartphone. Micromachines 2018:534. https://doi.org/10.3390/mi9100534.

[30] Lin S-S, Lan C-W, Hsu H-Y, Chen S-T. Data analytics of a wearable device for heat stroke detection. Sensors 2018;18. https://doi.org/10.3390/s18124347.

[31] Renardi MB, Basjaruddin NC, Rakhman E. Securing electronic medical record in near field communication using advanced encryption standard (AES). Technol Health Care 2018;26:357—62.
[32] Wang F, Preininger A. AI in health: state of the art, challenges, and future directions. Yearb Med Inform 2019;28:16—26.
[33] Quadros T de, de Quadros T, Lazzaretti AE, Schneider FK. A movement decomposition and machine learning-based fall detection system using wrist wearable device. IEEE Sensor J 2018:5082—9. https://doi.org/10.1109/jsen.2018.2829815.
[34] Brunner D, Vasko R, Detka C. Muscle artifacts in the sleep EEG: automated detection and effect on all-night EEG power spectra. J Sleep Res 1996;5(3):55—164. Available: https://onlinelibrary.wiley.com/doi/abs/10.1046/j.1365-2869.1996.00009.x.
[35] Rohál'ová M, Sykacek P, Koskaand M, Dorffner G. Detection of the EEG artifacts by the means of the (extended) kalman filter. Meas Sci Rev 2001;1:59—62.
[36] Yan PZ, Wang F, Kwok N, Allen BB, Keros S, Grinspan Z. Automated spectrographic seizure detection using convolutional neural networks. Seizure 2019;71:124—31.
[37] Liang J, Lu R, Zhang C, Wang F. Predicting seizures from electroencephalography recordings: a knowledge transfer strategy. In: 2016 IEEE international conference on healthcare informatics (ICHI); 2016. p. 184—91.
[38] Motamedi-Fakhr S, Moshrefi-Torbati M, Hill M, Hill CM, White PR. Signal processing techniques applied to human sleep EEG signals—a review. Biomed Signal Process Contr 2014;10:21—33.
[39] Altman NS. An introduction to kernel and nearest-neighbor nonparametric regression. Am Statistician 1992:175—85. https://doi.org/10.1080/00031305.1992.10475879.
[40] Liu K-C, Hsieh C-Y, Hsu SJ-P, Chan C-T. Impact of sampling rate on wearable-based fall detection systems based on machine learning models. IEEE Sensor J 2018:9882—90. https://doi.org/10.1109/jsen.2018.2872835.
[41] Hsieh C-Y, Huang C-N, Liu K-C, Chu W-C, Chan C-T. A machine learning approach to fall detection algorithm using wearable sensor. In: 2016 international conference on advanced materials for science and engineering (ICAMSE); 2016. https://doi.org/10.1109/icamse.2016.7840209.
[42] Menard S. Applied logistic regression analysis. SAGE; 2002.
[43] Muaremi A, Arnrich B, Tröster G. Towards measuring stress with smartphones and wearable devices during workday and sleep. BioNanoScience 2013:172—83. https://doi.org/10.1007/s12668-013-0089-2.
[44] Page A, Pramod Tim Oates S, Mohsenin T. An ultra low power feature extraction and classification system for wearable seizure detection. Conf Proc IEEE Eng Med Biol Soc 2015;2015:7111—4.
[45] Kamiński B, Jakubczyk M, Szufel P. A framework for sensitivity analysis of decision trees. CEJOR Cent Eur J Oper Res 2018;26:135—59.
[46] Zhang T, Tang W, Sazonov ES. Classification of posture and activities by using decision trees. Conf Proc IEEE Eng Med Biol Soc 2012;2012:4353—6.
[47] Chen M, Yang J, Zhou J, Hao Y, Zhang J, Youn C-H. 5G-Smart diabetes: toward personalized diabetes diagnosis with healthcare big data clouds. IEEE Commun Mag 2018:16—23. https://doi.org/10.1109/mcom.2018.1700788.
[48] Bellos CC, Papadopoulos A, Rosso R, Fotiadis DI. Identification of COPD patients' health status using an intelligent system in the CHRONIOUS wearable platform. IEEE J Biomed Health Inform 2014:731—8. https://doi.org/10.1109/jbhi.2013.2293172.
[49] Support Vector Machine. Support vector machine in chemistry. 2004. p. 24—52. https://doi.org/10.1142/9789812794710_0002.
[50] Kumar S, Nilsen W, Pavel M, Srivastava M. Mobile health: revolutionizing healthcare through transdisciplinary research. Computer 2013;46:28—35.

[51] Banaee H, Ahmed MU, Loutfi A. Data mining for wearable sensors in health monitoring systems: a review of recent trends and challenges. Sensors 2013;13:17472−500.

[52] Alpaydin E. Introduction to machine learning. MIT Press; 2014.

[53] Sreejith S, Rahul S, Jisha RC. A real time patient monitoring system for heart disease prediction using random forest algorithm. In: Advances in signal processing and intelligent recognition systems. Springer International Publishing; 2016. p. 485−500.

[54] Harris A, True H, Hu Z, Cho J, Fell N, Sartipi M. Fall recognition using wearable technologies and machine learning algorithms. In: 2016 IEEE international conference on big data (big data); 2016. p. 3974−6.

[55] Hastie T, Tibshirani R, Friedman J. The elements of statistical learning: data mining, inference, and prediction. Springer Science & Business Media; 2013.

[56] Mahmud MS, Fang H, Wang H, Carreiro S, Boyer E. Automatic detection of opioid intake using wearable biosensor. Int Conf Comput Netw Commun 2018;2018:784−8.

[57] LeCun Y, Bengio Y, Hinton G. Deep learning. Nature 2015;521:436.

[58] Aggarwal CC. Restricted Boltzmann machines. Neural networks and deep learning. 2018. p. 235−70. https://doi.org/10.1007/978-3-319-94463-0_6.

[59] Hinton G. Boltzmann machines. In: Encyclopedia of machine learning and data mining; 2014. p. 1−7. https://doi.org/10.1007/978-1-4899-7502-7_31-1.

[60] Miotto R, Wang F, Wang S, Jiang X, Dudley JT. Deep learning for healthcare: review, opportunities and challenges. Briefings Bioinf 2018;19:1236−46.

[61] Jindal V, Birjandtalab J, Pouyan MB, Nourani M. An adaptive deep learning approach for PPG-based identification. Conf Proc IEEE Eng Med Biol Soc 2016;2016:6401−4.

[62] Schmidhuber J. Deep learning in neural networks: an overview. Neural Network 2015:85−117. https://doi.org/10.1016/j.neunet.2014.09.003.

[63] Khan SS, Taati B. Detecting unseen falls from wearable devices using channel-wise ensemble of autoencoders. Expert Syst Appl 2017;87:280−90.

[64] Gu F, Khoshelham K, Valaee S, Shang J, Zhang R. Locomotion activity recognition using stacked denoising autoencoders. IEEE Internet Things J 2018;5:2085−93.

[65] Lecun Y, Bottou L, Bengio Y, Haffner P. Gradient-based learning applied to document recognition. Proc IEEE 1998:2278−324. https://doi.org/10.1109/5.726791.

[66] Esteva A, Kuprel B, Novoa RA, Ko J, Swetter SM, Blau HM, et al. Corrigendum: dermatologist-level classification of skin cancer with deep neural networks. Nature 2017;546:686.

[67] Zhang H, Wang A, Li D, Xu W. DeepVoice: a voiceprint-based mobile health framework for Parkinson's disease identification. In: 2018 IEEE EMBS international conference on biomedical health informatics (BHI); 2018. p. 214−7.

[68] Goodfellow I, Bengio Y, Courville A. Deep learning. MIT Press; 2016.

[69] Quisel T, Foschini L, Signorini A, Kale DC. Collecting and analyzing millions of mhealth data streams. Proc 23rd ACM 2017:1971−80. Available: https://dl.acm.org/doi/abs/10.1145/3097983.3098201?casa_token=g9Bk4xpuL4MAAAAA:PucaVbjnh_vwprjOk_BCzAy8Z7Faeqwfnfy3xjAqEfLl1KrCre_n0cIUA9gKCzf0DnUM5WsQ81L48Q.

[70] Wang F, Casalino LP, Khullar D. Deep learning in medicine—promise, progress, and challenges. JAMA Int Med 2019:293. https://doi.org/10.1001/jamainternmed.2018.7117.

[71] Wang F, Kaushal R, Khullar D. Should health care demand interpretable Artificial intelligence or accept "black box" medicine? Ann Intern Med 2020:59−60. Available: https://annals.org/acp/content_public/journal/aim/938321/aime202001070-m192548.pdf?casa_token=txIuJx8CSTIAAAAA:OKUGKuPgjeFXJTQUxXFiXo_dfJau8BFvQpyZBJctNPl4MwYShu64qF0QIp0WPh3eCxEzZ65NCnU.

[72] Oliver N, Lepri B, Sterly H, Lambiotte R, Deletaille S, De Nadai M, et al. Mobile phone data for informing public health actions across the COVID-19 pandemic life cycle. Sci Adv 2020:eabc0764. https://doi.org/10.1126/sciadv.abc0764.

[73] Wong CK, Ho DTY, Tam AR, Zhou M, Lau YM, Tang MOY, et al. Artificial intelligence mobile health platform for early detection of COVID-19 in quarantine subjects using a wearable biosensor: protocol for a randomised controlled trial. BMJ Open 2020:e038555. https://doi.org/10.1136/bmjopen-2020-038555.

[74] Huckins JF, Dasilva A, Wang W, Hedlund EL, Rogers C, Nepal SK, et al. Mental health and behavior during the early phases of the COVID-19 pandemic: a longitudinal mobile smartphone and ecological momentary assessment study in college students. J Med Int Res 2020. https://doi.org/10.31234/osf.io/4enzm.

[75] Sun S, Folarin A, Ranjan Y, Rashid Z, Conde P, Stewart C, et al. Using smartphones and wearable devices to monitor behavioural changes during COVID-19. 2020. arXiv [q-bio.QM]. Available: http://arxiv.org/abs/2004.14331.

CHAPTER 5

mHealth for research: participatory research applications to gain disease insights

Ipek Ensari, PhD [1], **Noémie Elhadad, PhD** [1,2]
[1]Data Science Institute, Columbia University, New York, NY, United States; [2]Department of Biomedical Informatics, Columbia University, New York, NY, United States

Contents

Learning objectives

Through this chapter, readers will be able to

- Evaluate contribution of self-tracking as participatory research activity in the context of mHealth research
- Analyze key health-informatics components for assessing mHealth application user engagement
- Identify techniques to gain insights from self-tracked data
- Integrate concepts of participatory research, user engagement, and health informatics in the context of mHealth and chronic diseases

Digital Health
ISBN 978-0-12-820077-3
https://doi.org/10.1016/B978-0-12-820077-3.00005-5
79

1. Background on research mHealth applications

Patient/person-generated health data (PGHD) is defined by the US Department of Health and Human Services as *"any health-related data, including health history, symptoms, biometrics, treatment history, lifestyle choices, and other information that is created, recorded, gathered, or inferred by or from persons or their designees to help address a health concern"* [7]. PGHD can be generated from a wide variety of sources including smartphone apps, wearable devices, and web-based patient portals and communities. Self-tracking as a participatory research activity to generate new knowledge from shared real-world experiences (e.g., *Patients Like Me* platform [8,9]) is becoming increasingly central to biomedical research, making it a suitable first step toward better understanding diseases [10] via active patient participation in citizen science [11]. As such, self-tracking of one's health and disease patterns to produce PGHD provides a promising approach to engage patients in disease self-management and self-discovery [12—15]. Participatory personal data in this context can be defined as any representation recorded by an individual, about an individual, using a mediating technology (i.e., mobile device) [16]. Recent clinical research, aided in part by the release of the Apple ResearchKit framework [17], has started leveraging self-tracking toward understanding diseases at scale [1,18,19] and has been shown to have advantages over traditional approaches for reaching to participants of diverse characteristics and ability to collect much larger magnitude of data [3].

Yet, emerging evidence reports substantial lack of engagement among participants of mHealth studies [5,20—22], which suggests a research gap in our knowledge about recruitment, engagement, and retention strategies for mHealth research interventions. This is an important point of inquiry as user engagement is a significant predictor of retention [5] and therefore sustained data collection [4], which are needed for gaining meaningful insights from these data and making accurate disease discoveries. Furthermore, despite the increasing number of research applications (apps) being made available to the public, there still lacks evidence on which design and mHealth functionalities are the most effective for user engagement and retainment beyond initial recruitment. Addressing these issues can maximize the opportunities of mHealth research to better understand how self-tracking data may be leveraged to create new knowledge about disease beyond personal discovery, and such opportunity is particularly significant when it comes to conditions that remain enigmatic [10,23].

In this chapter, we use a case study to guide the reader through the steps of making informed decisions on designing an mHealth research app for maximizing engagement and retention, metrics, and approaches to assessing these factors and, finally, demonstrate through an example how to generate disease insights from such data.

1.1 A case study: "The Citizen Endo Project and the Phendo Research App"

Citizen Endo is a research project[1] aimed to better understand endometriosis through direct patient input. Endometriosis is a systemic, estrogen-dependent inflammatory condition which causes chronic, noncyclical body pain [24]. It is estimated to affect 6%—10% of women of reproductive age, with a peak incidence between ages 25—30 years [25]. Chronic pelvic pain is the most frequent symptom; others include dyspareunia, dysuria, ovulation pain, and dysmenorrhea [25]. These impact on physical, social, and psychological function and significantly impair quality of life [26,27]. The large heterogeneity of the symptoms across individuals and within-individual temporal fluctuations result in an average delay of $6.7(\pm 6.3)$ years between symptom onset and surgical diagnosis [28], and in a disconnect between how healthcare professionals conceptualize endometriosis versus how patients experience it. Due to the large gap in the medical literature regarding this disease and lack of cure or effective treatment options, it is considered an "enigmatic" disease: poorly understood and understudied.

2. User engagement and participatory design in mHealth research

Well-designed mHealth apps have the potential to engage patients as advocates in their personalized care, offer healthcare providers real-world assessments of their patients' disease patterns, and better understand and gain insights into understudied diseases about which researchers and clinicians know little about. This potential can be realized if input from the chronic disease patients is collected during the design of the mHealth tool [29,30], specifically to assess what factors can increase the personal relevance of such a tool and how to ensure adequate usability and acceptability. This can contribute substantively to sustained user retention and engagement, which in turn enables accumulation

---

[1] http://citizenendo.org/.

of the necessary data to gain meaningful disease insights. As such, consideration of user engagement should start at the design stage and continue through deployment (i.e., active usage and data collection).

Defining Engagement. Although various definitions of engagement exist in the literature [21], for the purposes of this chapter, it can be conceptualized as usage of mHealth technologies in terms of temporal patterns (e.g., frequency, duration) and depth (e.g., use of specific intervention content) [21]. In other words, it is the degree of interaction a user has with the technology within a given time span or the overall length of time from the onset use of a technology to when the user totally lost interest in the usage [20]. It is a dynamic process that varies both within and across individuals; users can engage, disengage, and reengage with the mHealth technology over time [4,15].

Factors reported to influence engagement include those at the population level user-specific characteristics (i.e., physical, psychological, cognitive, demographic, mental health), at the user level (user expectations and personal relevance of the mHealth technology), and at the mHealth app level (e.g., app content such as its features with social support and reminders, app's delivery mode) [6,31]. Given the reported low retention of usership and short-lasting engagement with mHealth technologies [4,32], leveraging what is known about these factors and when to use them can help target the right ones during the design stage, therefore increasing likelihood of sustained engagement.

One limitation in the existing literature investigating the link between engagement and mHealth app design, however, is its reliance on interventions (e.g., behavioral change interventions) (e.g. Refs. [2,13—15]). **There are no guidelines on design of mHealth apps where the goal is to gain insights about the disease population**, instead of "intervene" on an individual's behavior. Particularly for conditions that are not well understood, a critical question is how to design such tools when it is unclear which data types are relevant to the disease. Because the successful use of an mHealth app requires an ability to understand and utilize personal health data, user experience should account for individual differences in numeracy skills and apply evidence-based behavioral science principles to promote continued engagement [33]. These collectively underscore the value of a patient-centered, participatory design approach in the context of observational research mHealth apps.

Stages of Participatory Design. Participatory design is a multistage, multi-method process [23,30,34], ideally guided through questions that the research team has set prior to beginning their research process (see Table 5.1).

Table 5.1 Guiding questions for conducting research to develop observational research app.

Steps	Guiding questions	Measurement approach and method
Determine target demographic group/disease and potential benefits	• Will this group benefit from an mHealth approach to better understanding the disease? • Is this group a mobile technology using group? • What are the researcher's goal in studying this group using mHealth technology?	• Demographic research on mobile technology usership (quantitative: structured surveys) • Assessment of interest in specific variables to track from the perspective of the target group (qualitattive: semistructured interviews and focus groups)
Identify key concerns of the target group for tracking	• What matters to the members of the demographic with respect to their disease? • What are they curious about and wanting to track? What is their need?	• Dimensions/main tenets relevant to the experience of the condition according to patients (quantitative: surveys, qualitative: semistructured 1 to 1 interviews, focus groups)
Identify additional dimensions of disease	• What further information could be extracted from already existing sources on the disease? • Which variables identified thus far inform the design of subsequent methods and steps of the project? • What metrics will be included in the user analytics component of the app?	• Quantitative; content analysis, surveys
Design iteratively with participants	• Are the selected dimensions meaningful to participants? • What is the framing, look, and feel of the app (e.g., scientific, coaching)?	• Mixed-methods; card-sorting experiments, face-to-face interview with participants, online survey, beta-testing of the app

Typically, as a first step, individual interviews and/or focus groups are conducted to gain an understanding of the needs, goals, and preferences of the target demographic. This qualitative phase can be supplemented with quantitative data collection approaches (e.g., surveys and questionnaires) to gain a comprehensive profile of the target user group. The design and the content of the mHealth tool are informed by the triangulation of data collected through these stages and in an iterative manner tested with groups of participants. This iteration phase includes usability, acceptability, and learnability assessments, conducted at multiple stages as needed during the design process. Importance of this iterative stage is underscored by the significant association of these three factors to research mHealth tool user retention and engagement [30,32].

From a quantitative approach, there is a variety of models and validated questionnaires for assessing these components, including the Health IT Usability Evaluation Model (Health-ITUEM) [35], Technology Acceptance Model (TAM), the customizable Health IT Usability Scale (Health-ITUES) [36] that was developed based on the framework, IBM Computer System Usability Questionnaire [37], System Usability Scale [38], End-User Computing Satisfaction [39], and Unified Theory of Acceptance and Use of Technology (UTAUT) [40]. Information gained through the structured questionnaires can further be supplemented with qualitative assessment of user perceptions of the design through focus groups and individual interviews.

In sum, this multistage data triangulation process helps circumvent limitations of digitization of chronic diseases identified in the literature, including lack of consumer appeal or clinical relevance, low added value, missing services for comorbidities, and demographic and behavioral characteristics of patient populations [2]. As such, it is recommended that these potential issues are considered as early as possible in the app design process.

2.1 Case study: participatory design of Phendo

Phendo is an observational research smartphone app available for iOS and Android, and was developed for people with endometriosis to self-track symptoms, treatments, and self-management strategies associated with their disease [11,23]. The app enables participants to review and sign an electronic consent to participate to the Citizen Endo research. In terms of self-tracking features within Phendo, participants can track the location, intensity, and type of their endometriosis pain across 39 specific body locations, gastrointestinal and genitourinary issues, other symptoms (e.g., "blurry vision" "hot flashes" "fatigue"), and their severity, mood, bleeding, and medication intake. Users can track a functional assessment of their day ("How was your day?"), menstruation patterns, and a daily journal. All aspects of Phendo were designed based on a series of quantitative and qualitative exploratory research conducted with women living with endometriosis [11], using a multistep data triangulation process described above. The goal of the research team was to create **a patient-centered tool that engages the user as an active participant in the research on better understanding endometriosis**. Further details of each of these steps are provided in Ref. [23] and we summarize the guiding questions that informed each step as a guiding checklist for the reader (see Table 5.1). Below, we describe some of the key learnings gained at each step.

Initially, the research team conducted a round of individual interviews with three women with endometriosis to collect information on personal history of the disease for each woman, their symptoms, and the burden of disease on their everyday lives. This was followed by a series of five focus groups held with women (N = 27, age ranging between 27 and 60 years) with endometriosis to (1) explore their attitudes and motivations about self-tracking their experience of the disease and (2) determine the feasibility of self-tracking endometriosis. The results from these focus groups revealed several important points, which informed the subsequent stages of the Phendo project. First, we discovered that women living with endometriosis were eager to self-track as a form of participatory research activity and that they believed their input could help clinicians and researchers better understand this disease [11]. Next, these interviews and focus groups identified aspects of the disease they were willing and thought of as important to track (i.e., prioritized disease aspects) [23].

Based on the findings from the focus groups, the research team **developed questionnaires focusing on disease variables discussed during these sessions**, including emotions, moods, and affects; pain locations and descriptions; medication use; and self-management, strategies, comorbidities, dietary, and exercise habits. Importantly, participants were informed of the explicit goal of designing a self-tracking tool (i.e., participatory design). Finally, to further explore the dimensions identified during the prior stages without interference from the potential biases of the researchers, a content analysis of an online health community (i.e., the endometriosis board, "r/endo," of the social platform Reddit) was conducted. This content analysis provided confirmation of the disease aspects identified in the focus groups and online surveys [23]. This approach further enables capturing of shared, real-life experiences of individuals with the disease and can be helpful in instances where not much is known about the disease.

Included within the Phendo app user tab, in addition to researcher-derived items, Phendo participants can take a standardized questionnaire designed by the World Endometriosis Research Foundation [26] Endometriosis Phenome and Biobanking Harmonisation Project (WERF-EPHect) [41]. The questionnaire represents the gold standard for clinical characterization of endometriosis. It contains information about medical and surgical history, as well as quality of life-related questions. Outside of endometriosis, other examples of standardized, validated measures that can be used across a multitude of conditions are the patient-reported outcome measures (PROMs) and the patient-reported experience measures (PREMs). A PRO is directly reported by the patient without interpretation of the patient's response by a clinician or anyone else, and pertains to the patient's health, quality of life, or functional status associated with healthcare or treatment [42]. Similarly, PREM is defined as a measure of a patient's perception of their personal experience of the healthcare they have received [43]. In the context of observational mHealth research, inclusion of standardized measures allows for **combining self-tracked mobile data with those from gold standard measures for comprehensive profiling of the disease, and gain further understanding of lesser-known disease aspects** (see below in Section 5).

3. User engagement strategies and metrics

mHealth User Engagement Strategies. Engagement strategies should be determined based on the research question and goal of the researcher with respect to the app usage, while keeping in mind the within- and between-user heterogeneity in the pattern of engagement that occurs naturally over time [44]. In the case of self-tracking as a participatory research activity (as opposed to partake in the app to change a behavior), a main goal is to recruit, engage, and retain users to track data on participation in their self-care activities. Participation in these activities may be reflected in the frequency of tracking or monitoring of the activities using an app hence denoting the extent of their engagement with the app [4,5]. If a user can interact with the app as often as they like, another measure of long-term engagement is the number of interactions—or the percent of recommended interactions—they conduct within a given time span [20]. A caveat is nonusage, which is that patients may participate in self-care activities and not track with an app even when apps are present on their mobile devices [45], which may be due to reduced perceived quality of experience and ongoing benefit of usage, consideration of viable alternatives to using the technology, and decrease in perceived costs. In sum, apps offer motivation and support for self-management of various disease conditions (e.g. Ref. [46]), but their use is limited by high attrition likely due to insufficient consideration of end-users' perspectives and usability requirements. As such, identification of feature preferences among the target user demographic and their overall investment in the mobile app as a first step is helpful to foster engagement with the created app [5,47].

Engagement Strategies for Observational mHealth Research. For observational research apps, engagement strategies differ on some levels in comparison to those for behavior change apps, and focus on recruiting, retaining, and engaging participants (in contrast, for example, a behavior change app also needs to include a metric for assessing whether the target behavior change has occurred and has been sustained). As one example, regular usage of the self-monitoring function has been identified as a significant predictor of sustained use of mHealth apps [48], and might suggest that focusing on a functional, easy to use self-monitoring feature within the app should be a priority when designing apps for observational research.

Gamification, use of game design elements in nongame contexts [49], has also been studied as an approach for engaging mHealth app users [49], mainly across intervention-based mHealth apps [50]. This approach has potential for improving mood [50,51] and health behaviors such as physical activity [52], smoking cessation [53], and diabetes management [54]. However, there is inadequate evidence on its application in the context of observational studies [18,56] or effectiveness for long-term engagement [50]. Accordingly, further research is needed to better understand how gamification can improve participant mHealth app use in the long term [50,51].

With respect to long-term engagement, an analysis [44] of eight remote digital health studies comprising over 100,000 participants who collectively contributed more than 850,000 days of track data and 3.5 million remote health evaluations concluded that the most influential predictors for retaining participants over time include clinician referral of the participant to the study, monetary compensation through use of "pay-for-participation" models, and inclusion of clinical populations and older adults.

Another suggested strategy is to tailor the components of the study, including enrollment, in-app communication, and return of information to the participants, based on the distinct characteristics of the three types of technology adopters: trendsetters (early adopters), majority users, and laggards (highly resistant to change and hard to reach online) [44].

3.1 Case study: engagement strategies explored in Phendo

Through our team's ongoing work in Phendo, here we share different strategies implemented for recruiting, engaging, and retaining users, specifically tailored for observational research (see Table 5.2).

In the beginning stages of Phendo, participants were recruited through patient advocacy groups and, once enrolled, can self-track a variety of variables. Later, recruitment happened through patient meet-ups, community outreach events, and online advertising through Citizen Endo's twitter and medium blog. Currently, strategies our team relies on for recruiting participants and new users for Phendo include use of social media, where we use our regularly updated medium blog page, Instagram and twitter accounts to share updates and news regarding participation opportunities.

Once recruited, the focus shifts to retaining the users and engaging them to contribute data on their disease. To accomplish this, we use various tactics (see Table 5.2), including use of social media. While the use of social media in the context of health promotion has been on the rise [57], it has been limited to behavior change interventions [55]. One strategy we have

Table 5.2 Recruitment and engagement strategies explored in Phendo.

Recruitment	Retainment/Engagement
- Social media, patient meet-ups, news and patient advocacy groups, popular media figures as patients - Participant self-tracking challenges - Citizen science	- App functionalities for retroactive self-tracking in time and ability to edit self-tracking events - Participant self-tracking challenges - Citizen science - Self-management feature within the app (for facilitating communication with provider, and reviewing trends in disease experience) - Customization and personalization at the disease level (tracking, visualizations)

observed to be helpful are occasional user challenges within the app ("citizen science challenges"), which involve having short-term (e.g., 14 days) data collection periods during which volunteering Phendo users track a given a set of variables every day specific to a theme (e.g., physical activity patterns, disease symptoms) (Fig. 5.1). These challenges not only capture the interest of the users but also provide a novelty component within the app, which prevents boredom or loss of interest over time. After the data collection challenge period, we share aggregated, deidentified results with the participants on the projects' social media blog, as a way to

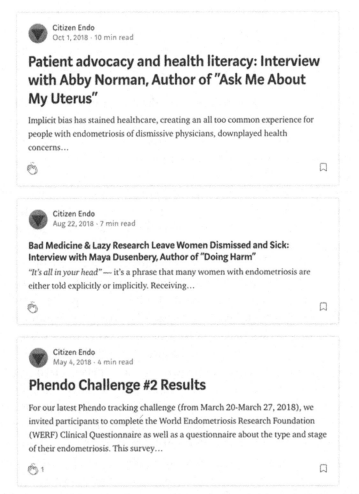

Figure 5.1 Examples of Phendo social media posts with various content, including patient advocacy and health literacy, expert interviews, and citizen science tracking challenges.

return results to the participants. Similarly, we keep an active online presence on our social media pages (see Fig. 5.1), and post links to events relevant to our demographic (e.g., patient days), news or updates on our research group, as well as within-app user trends (e.g., most frequently tracked variables). These updates and links help create a sense of community for the users, which further nurture the "citizen science" aspect of Citizen Endo that participants are contributing information about their disease as a participatory research activity.

Also relevant to observational mHealth apps, personalization and customization are two methods of tailoring systems for retainment and engagement. Customization indicates user-controlled tailoring, while personalization is mostly system controlled, and both have been reported to be effective depending on the context [6,31,56]. Customization further has been associated with increased user sense of personal agency and control [31]. Similarly, provision of means for users to view and compare their data to those of others is another strategy for the same purpose [6], and this can be accomplished via integrating features into the app, for example, via visuals that are customizable (e.g., choosing symptoms from the drop-down menu, creating multiple graphs for comparison). Within the Phendo app, users are able to customize their tracking of which activities of daily living were hard to do, dietary habits, supplements, exercises, and hormonal treatments. Customization involves the user adding these items as variables in the app and being able to save and select thereafter for tracking.

Visualizations for Personalized Insights. Through our team's ongoing work on self-management strategies in endometriosis and patient-centered design [57], we have created various features within the app for the participants to generate tracking history visualizations, both for their personal use and to share with their healthcare provider. Fig. 5.2 shows a screenshot from the Phendo app that allows the users to look at tracking history for multiple symptoms and self-management strategies that can be shared with their healthcare provider, as well as charts that generate personalized history of a chosen symptom. These were created through an iterative process whereby both the patient and healthcare provider give feedback on the design, content, and format of the visualizations from a usability and functionality perspective [57]. For further reading on these strategies, see [56].

It is important to note here the distinction between different *levels* of tailoring. In the context of Phendo, the aim was to better understand an enigmatic disease and as such the research team aimed to create **disease-level tailoring**, as opposed to customization at the individual level. It is

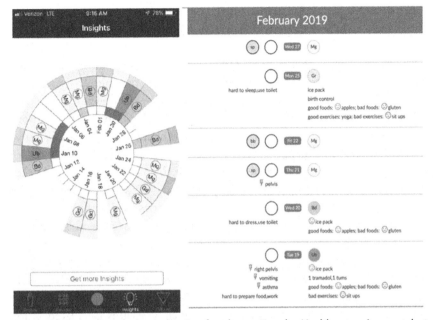

Figure 5.2 Examples of personalization for observational mHealth apps. A screenshot of the "Personalized Insights" feature within the Phendo app, which allows for generation of a tracking history graph for a symptom (left); and a screenshot of the desktop version of a symptom and self-management history report for sharing with participant's own healthcare provider (right).

still possible to integrate person-level customization, such as allowing the user to pick a certain disease-related variable and/or health behavior to self-monitor over time. Indeed, such self-monitoring ability within mHealth apps has been indicated to be a significant predictor of long-term, committed use of the technology [48], suggesting that users are motivated by tracking and updating their health information accurately so as to stay informed about the consequences of their daily activities, which can lead to healthier behavior and better clinical outcomes. In sum, the level of tailoring will depend on and should be driven by the research question and the motivation of the project.

Engagement Metrics for mHealth. Based on the strategies discussed, there are various analytic indicators for assessing effective real-time user engagement with consumer and research type mHealth apps reported in the literature [5,22] Any engagement metrics chosen for a given study should align with the strategies used for retaining and engagement strategies. We identify, in our experience with Phendo, traditional engagement metrics (e.g., retainment at a particular time) and engagement metrics specifically for disease insight discovery (e.g., span of time, diversity of self-tracked days), described below.

User Engagement Metrics Reported in the Literature. Though more recently recommendations and guidelines on mobile app usership analytics have been published [58,59], existing literature rather focuses on evaluating metrics for behavior change interventions [5,14,15] and includes (1) reports of the subjective user experience, elicited by qualitative methods or questionnaires, and (2) objective measures of technology usage, user behavior and physiological measures, users' reactions to the intervention, and efficacy and effectiveness outcomes [15,58,59]. Even so, still there are currently no published, standardized app usage frequency metrics used across mobile app studies, and particularly for observational apps. This has led some researchers to define their own criteria based on need and purpose of their study or borrow concepts from nonhealth-related fields (e.g., Refs. [47,60—62]) to adapt for their own needs. As one example, Cheung et al. [61] compute an *overall user engagement* index in their study, which is a composite score that integrates three different metrics. *Continued engagement* is calculated as the duration between downloading the app and the last use of any app, *user loyalty* is measured as the number of sessions that a user launched in a week over a set period (16 weeks in their case), and third, *user regularity* is measured as the number of days with at least one app session launched within a week over the set period (i.e., 16 weeks in the same example). These three metrics combined constitute the "overall user engagement" composite score, which provides a comprehensive profile of user engagement, reflects longitudinal trend of usage, and as such circumvents the miscount of instances where a user might have high weekly loyalty but low regularity if s/he used the app many times on a single day and at no times on other days. For further reading on metrics of user engagement in the context of behavioral interventions, see Ref. [15,58]. As a strategy, we use and recommend using **multiple measures of engagement to capture a comprehensive profile of user engagement, while being mindful of minimizing user burden** and measurement effects as far as possible.

User Engagement Metrics for Observational mHealth Research. Similar to abovementioned engagement strategies, metrics of engagement for observational mHealth apps differ on various levels in comparison to those for behavior change apps. For example, effectiveness and efficacy, adherence to intervention, or health outcomes are not the main focus of these apps, since the goal is not to change behavior or improve an outcome. Rather, the focus is on sustained, meaningful interaction with the app such that the user can provide rich contextual data that will lead to generation disease insights and discoveries by the research (see Section 5 for a case study). Accordingly, system usage data should be explored while keeping in mind the multidimensional nature of engagement, and focus on the breadth, depth, and type of engagement as well as frequency measures [14,63].

One such approach has been proposed by [14] in the context of behavior change apps and involves mapping the engagement measures on (1) frequency, (2) intensity, (3) time, and 3 type, borrowing the "FITT" acronym from the field of Kinesiology. This type of mapping can also apply to observational research apps. Accordingly, frequency metrics include those such as the number of visits or log-ins per participant per unit of time. Intensity metrics are those that are related to the depth of the participant's engagement and include those such as number of app components viewed (e.g., pages, modules), number of tracks, and number of components created within the app (e.g., number of customized health behaviors and disease symptoms created by the user). Duration metrics include amount of time spent at each app visit and range of days of use (i.e., first to last days of log in). Duration of stay per visit ("*stickiness*") has been associated with increased engagement [63,64], and evidence suggests that this is significantly influenced by participant attitude toward the content and its source [64], including whether the user considers the source to be a reliable authority on the subject or not [45]. Interestingly, this concept of stickiness and loyalty comes from the field of e-commerce [64,68], but has practical use in the context of mHealth content use.

From an implementation standpoint, the engagement metrics described above can be computed from self-tracking events. To assess the engagement of participants with the app itself, more traditional log data beyond self-tracking are useful and help answer questions such as how participants behave and how much time they spend interacting with the app. This type of engagement can be particularly informative to identify app features that are more or less popular with participants, especially for the ones that do not involve tracking. For instance, if a research app has an educational component, which simply displays information, keeping track of its usage and popularity among participants may better help assess and adjust the type and content of that functionality. Ultimately, which combination of metrics used should depend on the disease, research questions, and even within the same study, which metrics will be most meaningful might vary over time.

4. Making disease discoveries and insights from self-tracked data

Self-tracking through mHealth apps as a form of participatory research can enable researchers to gain clinically meaningful insights from these PGHD [65–68], which are particularly useful when studying diseases that are poorly understood and inadequately treated, as well as those that are cyclical and with symptoms fluctuating in time [66,67]. For example, Li et al. [67] demonstrate that menstrual cycle variability between individuals is associated with distinct clusters of patient symptom profiles, and this has implications for improving the diagnosis and prediction of menstrual disorders [67]. In contrast to self-tracking data, there are examples of large-scale studies that utilize passively generated data through mobile apps, where the contributor of the data is not actively involved in the research process. For example, Althoff et al. [69] leverage smartphone accelerometry data obtained from 717,527 individuals across 111 countries to measure the distribution of physical activity and its inequality within countries. Their findings indicate that this inequality is a better predictor of obesity than the volume of physical activity [69], and that the built environment (e.g., walkability of neighborhood) is linked to gender gaps in activity.

To date, there has been limited work systematically investigating how mHealth participatory research can be retooled to make discoveries. Here we discuss an example of machine learning–based modeling of Phendo data to derive clinically relevant subtypes ("phenotypes") of endometriosis, the interpretability of this information, and potential directions for closing the loop of transfer of knowledge by providing feedback to the participant based on their individual findings.

4.1 Case study: gaining insights from participatory research

As mentioned earlier, Phendo was built to better document and understand endometriosis and its symptom heterogeneity between patients and over time. This heterogeneity suggests that different phenotypes of endometriosis might exist, thus requiring more accurate phenotyping of the disease to inform the development of targeted and effective treatments and management strategies [41]. However, because endometriosis is not well understood from a clinical point of view, traditional phenotyping approaches based on EHR data are not appropriate.

To address this gap, Urteaga et al. [70] leveraged Phendo data to demonstrate that PGHD can be used to identify distinct phenotypes of endometriosis. An unsupervised mixed-membership modeling approach [71] was used for phenotyping to account for the heterogeneity in the disease experience across patients and also in the tracked mobile data (i.e., many variables available for self-tracking, not all variables are measured at the same time points from all users or at regular intervals; participants may track more often when they experience symptoms of disease or not self-track when they are sick). Each set of observations was modeled as a mixture model [72] and phenotypes were inferred based on the co-occurrence of observations across the studied set of participants, by creating groupings of responses to self-tracked variables that describe endometriosis phenotypes [70]. The quality of the learned phenotypes was evaluated by how they correlate with clinical expert disease groupings and with responses to a standardized endometriosis severity survey [70].

Four distinct phenotypes (i.e., 0, 1, 2, 3) were identified, each of which can be described through the app variables, making the phenotypes interpretable. For example, Phenotype 0 was identified as the most severe cluster, with reports of a much wider range of pain locations compared to the other phenotypes, and also more frequent genitourinary symptoms. While genitourinary symptoms in endometriosis are known, their association with a subgroup of patients is novel [73]. Phenotypes 1 and 2 were associated with pain primarily in pelvis, uterus, or vagina. The tracked pain was commonly described as aching or cramping across all phenotypes, while Phenotype 0 had a higher likelihood of deep pain reports, and was uniquely associated with burning, throbbing, and nauseating pain. Common across all phenotypes were psychological symptoms including fatigue, mental fogginess, and headache. This overarching symptomology is also prevalent in other chronic conditions (e.g., chronic fatigue syndrome [74], multiple sclerosis [75]) and has been linked to low grade inflammation [76]. Identification of disease subtypes with distinct characteristics demonstrates that we can produce scalable clinically meaningful groupings of self-tracked variables, which can be applied to other conditions beyond endometriosis. Finally, another important take-away from this case study is that it is imperative to have a plan in place for handling the noisy nature of the digital mHealth app data when setting out to conduct research, especially inferential work, using PGHD [67,70].

Future Directions for Participant Feedback. A possible next step would be to close this knowledge transfer loop by providing feedback to the participant, including actionable insights on what factors might be helpful for them to better manage their disease based on the characteristics of their phenotype.

Examples exists in the context of behavior change apps where approaches such as just-in-time interventions, N-of-1 designs, and artificial intelligence-based reinforcement learning models are implemented to learn patterns at the individual level and tailor the feedback given to the participant with the goal of improving targeted health behaviors (e.g. Ref. [77–79]). In the context of observational research, integrated app functions for direct communication with the participant have been infrequently reported in the literature, despite their promise [80,81]. Regular communications to encourage continued app usage to motivate user engagement can take the form of individual or group messages that remind the participants the value of their contribution of data to the research via tracking [81], or sharing insights obtained at the disease level from the PGHD via app's social media or in-app communication functionalities with the users, as described in the Phendo Case Study under Section 4.

5. Conclusion and discussion

Throughout this chapter, we demonstrate that a successful mHealth app, one that garners sustained active usership and thus leads to generating meaningful insights, starts with good design and that good design starts with input from the target user demographic. These steps together constitute a three-component forward-feeding loop of *participatory research design—user engagement—disease insights,* which is particularly relevant for mHealth research apps aimed for better understanding understudied conditions and patient populations. While there is much promise and activity for this research topic, there are still several open research questions, including how to quantify engagement and retention with respect to ability to discover disease insight and which engagement strategies are more likely to promote sustained engagement in research-based apps.

Before setting out to conduct an mHealth research study, assessing whether the target disease population is amenable for this methodology will significantly boost the chances of success to answer the research questions for which the app is built. Our findings indicated that women with endometriosis are a particularly suitable population to engage in participatory research given that they were very interested in self-tracking, felt their symptoms are often ignored or dismissed by physicians, and that they have questions about their disease that have not yet been answered through biomedical research. However, this might not be the case for every demographic or applicable to every condition. A mixed-methods approach to characterize the attitudes of the target population can ensure that the designed research app will have acceptable use and garner data collection. Extensive a priori exploratory research with the target user population to receive initial input, followed by continuous feedback at later stages, enables the design of the app to be relevant and personally meaningful for the target group. This initial exploratory research consequently allows for more effective engagement of its users, and further places a focus on personal relevance, personalization of features of the app, which is an essential ingredient for engagement and retention in mHealth app-based research [21,31].

Acknowledgments

Preparation of this work was in part supported by an award from National Library of Medicine (R01 LM013043).

References

[1] Zens M, Woias P, Suedkamp NP, Niemeyer P. "Back on track": a mobile app observational study using Apple's ResearchKit framework. JMIR mHealth & uHealth 2017;5(2). https://doi.org/10.2196/mhealth.6259. PubMed PMID: 28246069. e23-e.

[2] Ben-Zeev D, Schueller SM, Begale M, Duffecy J, Kane JM, Mohr DC. Strategies for mHealth research: lessons from 3 mobile intervention studies. Adm Pol Ment Health 2015;42(2):157−67.

[3] Bot BM, Suver C, Neto EC, Kellen M, Klein A, Bare C, et al. The mPower study, Parkinson disease mobile data collected using ResearchKit. Sci Data 2016;3:160011.

[4] Scherer EA, Ben-Zeev D, Li Z, Kane JM. Analyzing mHealth engagement: joint models for intensively collected user engagement data. JMIR mHealth & uHealth 2017;5(1):e1.

[5] Baumel A, Kane JM. Examining predictors of real-world user engagement with self-guided eHealth interventions: analysis of mobile apps and Websites using a novel dataset. J Med Internet Res 2018;20(12):e11491.

[6] Orji R, Vassileva J, Mandryk RL. Modeling the efficacy of persuasive strategies for different gamer types in serious games for health. User Model User Adapt Interact 2014;24(5):453−98. https://doi.org/10.1007/s11257-014-9149-8.

[7] Deering MJ, Siminerio E, Weinstein S. Issue brief: patient-generated health data and health IT. Off Natl Coord Health Inf Technol 2013:20.

[8] Okun S, Wicks P. DigitalMe: a journey towards personalized health and thriving. Biomed Eng Online 2018;17(1):119. https://doi.org/10.1186/s12938-018-0553-x.

[9] Okun S, Goodwin K. Building a learning health community: by the people, for the people. Learn Health Syst 2017;1(3):e10028. https://doi.org/10.1002/lrh2.10028.

[10] Identifying rare diseases from behavioural data: a machine learning approach. In: MacLeod H, Yang S, Oakes K, Connelly K, Natarajan S, editors. 2016 IEEE first international conference on connected health: applications, systems and engineering technologies (CHASE). IEEE; 2016.

[11] McKillop M, Voigt N, Schnall R, Elhadad N. Exploring self-tracking as a participatory research activity among women with endometriosis. J Particip Med 201;8:e17.

[12] Boundary negotiating artifacts in personal informatics: patient-provider collaboration with patient-generated data. In: Chung C-F, Dew K, Cole A, Zia J, Fogarty J,

Kientz JA, et al., editors. Proceedings of the 19th ACM conference on computer-supported cooperative work & social computing. ACM; 2016.

[13] Lyons EJ, Lewis ZH, Mayrsohn BG, Rowland JL. Behavior change techniques implemented in electronic lifestyle activity monitors: a systematic content analysis. J Med Internet Res 2014;16(8):e192. https://doi.org/10.2196/jmir.3469.

[14] Short CE, DeSmet A, Woods C, Williams SL, Maher C, Middelweerd A, et al. Measuring engagement in eHealth and mHealth behavior change interventions: viewpoint of methodologies. J Med Internet Res 2018;20(11):e292.

[15] Yardley L, Spring BJ, Riper H, Morrison LG, Crane DH, Curtis K, et al. Understanding and promoting effective engagement with digital behavior change interventions. Am J Prev Med 2016;51(5):833−42.

[16] Shilton K. Participatory personal data: an emerging research challenge for the information sciences. J Am Soc Inf Sci Technol 2012;63(10):1905−15.

[17] Jardine J, Fisher J, Carrick B. Apple's ResearchKit: smart data collection for the smartphone era? London, England: SAGE Publications Sage UK; 2015.

[18] Chan Y-FY, Wang P, Rogers L, Tignor N, Zweig M, Hershman SG, et al. The Asthma Mobile Health Study, a large-scale clinical observational study using ResearchKit. Nat Biotechnol 2017;35(4):354.

[19] McConnell MV, Shcherbina A, Pavlovic A, Homburger JR, Goldfeder RL, Waggot D, et al. Feasibility of obtaining measures of lifestyle from a smartphone app: the MyHeart counts cardiovascular health study. JAMA cardiol 2017;2(1):67−76.

[20] Bickmore T, Schulman D, Yin L. Maintaining engagement in long-term interventions with relational agents. Appl Artif Intell 2010;24(6):648−66.

[21] Perski O, Blandford A, West R, Michie S. Conceptualising engagement with digital behaviour change interventions: a systematic review using principles from critical interpretive synthesis. Transl Behav Med 2016;7(2):254−67.

[22] Pham Q, Graham G, Carrion C, Morita PP, Seto E, Stinson JN, et al. A library of analytic indicators to evaluate effective engagement with consumer mHealth apps for chronic conditions: scoping review. JMIR mHealth & uHealth 2019;7(1):e11941.

[23] Designing in the dark: eliciting self-tracking dimensions for understanding enigmatic disease. In: McKillop M, Mamykina L, Elhadad N, editors. Proceedings of the 2018 CHI conference on human factors in computing systems. ACM; 2018.

[24] Tamaresis JS, Irwin JC, Goldfien GA, Rabban JT, Burney RO, Nezhat C, et al. Molecular classification of endometriosis and disease stage using high-dimensional genomic data. Endocrinology 2014;155(12):4986−99. https://doi.org/10.1210/en.2014-1490. PubMed PMID: 25243856; PubMed Central PMCID: PMCPMC4239429. Epub 2014/09/23.

[25] Rogers PA, D'Hooghe TM, Fazleabas A, Gargett CE, Giudice LC, Montgomery GW, et al. Priorities for endometriosis research: recommendations from an international consensus workshop. Reprod Sci 2009;16(4):335−46. https://doi.org/10.1177/1933719108330568. PubMed PMID: 19196878; PubMed Central PMCID: PMCPMC3682634.Epub 2009/02/07.

[26] De Graaff AA, D'Hooghe TM, Dunselman GA, Dirksen CD, Hummelshoj L, Consortium WE, et al. The significant effect of endometriosis on physical, mental and social wellbeing: results from an international cross-sectional survey. Hum Reprod 2013;28(10):2677−85. https://doi.org/10.1093/humrep/det284. PubMed PMID: 23847114.Epub 2013/07/13.

[27] Simoens S, Dunselman G, Dirksen C, Hummelshoj L, Bokor A, Brandes I, et al. The burden of endometriosis: costs and quality of life of women with endometriosis and treated in referral centres. Hum Reprod 2012;27(5):1292−9. https://doi.org/10.1093/humrep/des073. PubMed PMID: 22422778.Epub 2012/03/17.

[28] Nnoaham KE, Hummelshoj L, Webster P, d'Hooghe T, de Cicco Nardone F, de Cicco Nardone C, et al. Impact of endometriosis on quality of life and work productivity: a multicenter study across ten countries. Fertil Steril 2011;96(2):366–73. https://doi.org/10.1016/j.fertnstert.2011.05.090. PubMed PMID: 21718982; PubMed Central PMCID: PMCPMC3679489.e8. Epub 2011/07/02.

[29] Chiauzzi E, Rodarte C, DasMahapatra P. Patient-centered activity monitoring in the self-management of chronic health conditions. BMC Med 2015;13:77. https://doi.org/10.1186/s12916-015-0319-2. PubMed PMID: 25889598.

[30] Birnbaum F, Lewis D, Rosen RK, Ranney ML. Patient engagement and the design of digital health. Acad Emerg Med 2015;22(6):754–6. https://doi.org/10.1111/acem.12692. PubMed PMID: 25997375.Epub 2015/05/21.

[31] Towards personality-driven persuasive health games and gamified systems. In: Orji R, Nacke LE, Di Marco C, editors. Proceedings of the 2017 CHI conference on human factors in computing systems. ACM; 2017.

[32] Holden RJ, Karsh B-T. The technology acceptance model: its past and its future in health care. J Biomed Inf 2010;43(1):159–72. https://doi.org/10.1016/j.jbi.2009.07.002. PubMed PMID: 19615467.Epub 2009/07/15.

[33] Coughlin S, Thind H, Liu B, Champagne N, Jacobs M, Massey RI. Mobile phone apps for preventing cancer through educational and behavioral interventions: state of the Art and remaining challenges. JMIR mhealth & uhealth 2016;4(2):e69. https://doi.org/10.2196/mhealth.5361.

[34] Schnall R, Rojas M, Bakken S, Brown W, Carballo-Dieguez A, Carry M, et al. A user-centered model for designing consumer mobile health (mHealth) applications (apps). J Biomed Inf 2016;60:243–51.

[35] Brown III W, Yen P-Y, Rojas M, Schnall R. Assessment of the health IT usability evaluation model (Health-ITUEM) for evaluating mobile health (mHealth) technology. J Biomed Inf 2013;46(6):1080–7.

[36] Yen P-Y, Wantland D, Bakken S. Development of a customizable health IT usability evaluation scale. AMIA Annu Symp Proc 2010;2010:917–21. PubMed PMID: 21347112.

[37] Lewis JR. IBM computer usability satisfaction questionnaires: psychometric evaluation and instructions for use. Int J Hum Comput Interact 1995;7(1):57–78.

[38] The factor structure of the system usability scale. In: Lewis JR, Sauro J, editors. International conference on human centered design. Springer; 2009.

[39] Doll WJ, Xia W, Torkzadeh G. A confirmatory factor analysis of the end-user computing satisfaction instrument. MIS Quarterly 1994:453–61.

[40] Venkatesh V, Morris MG, Davis GB, Davis FD. User acceptance of information technology: toward a unified view. MIS Quarterly 2003:425–78.

[41] Vitonis AF, Vincent K, Rahmioglu N, Fassbender A, Louis GMB, Hummelshoj L, et al. World Endometriosis Research Foundation Endometriosis Phenome and bio-banking harmonization project: II. Clinical and covariate phenotype data collection in endometriosis research. Fertil Steril 2014;102(5):1223–32.

[42] Higgins JP, Green S. Cochrane handbook for systematic reviews of interventions. John Wiley & Sons; 2011.

[43] Male L, Noble A, Atkinson J, Marson T. Measuring patient experience: a systematic review to evaluate psychometric properties of patient reported experience measures (PREMs) for emergency care service provision. Int J Qual Health Care 2017;29(3):314–26. https://doi.org/10.1093/intqhc/mzx027.

[44] Pratap A, Neto EC, Snyder P, Stepnowsky C, Elhadad N, Grant D, et al. Indicators of retention in remote digital health studies: a cross-study evaluation of 100,000 participants. NPJ Digit Med 2020;3:1–10.

[45] Kayyali R, Peletidi A, Ismail M, Hashim Z, Bandeira P, Bonnah J. Awareness and use of mHealth apps: a study from England. Pharmacy 2017;5(2):33.

[46] Adu MD, Malabu UH, Malau-Aduli AEO, Malau-Aduli BS. Users' preferences and design recommendations to promote engagements with mobile apps for diabetes self-management: multi-national perspectives. PloS One 2018;13(12). https://doi.org/10.1371/journal.pone.0208942. PubMed PMID: 30532235; PubMed Central PMCID: PMCPMC6287843.e0208942. Epub 2018/12/12.

[47] Peterson ET, Carrabis J. Measuring the immeasurable: visitor engagement. Web Anal Demystified 2008;14:16.

[48] Lee K, Kwon H, Lee B, Lee G, Lee JH, Park YR, et al. Effect of self-monitoring on long-term patient engagement with mobile health applications. PloS One 2018;13(7):e0201166.

[49] Chatzitofis A, Monaghan D, Mitchell E, Honohan F, Zarpalas D, O'Connor NE, et al. HeartHealth: a cardiovascular disease home-based rehabilitation system. Procedia Comput Sci 2015;63:340—7.

[50] Sardi L, Idri A, Fernández-Alemán JL. A systematic review of gamification in e-Health. J Biomed Inf 2017;71:31—48. https://doi.org/10.1016/j.jbi.2017.05.011.

[51] A review of gamification for health-related contexts. In: Pereira P, Duarte E, Rebelo F, Noriega P, editors. International conference of design, user experience, and usability. Springer; 2014.

[52] Garde A, Umedaly A, Abulnaga SM, Robertson L, Junker A, Chanoine JP, et al. Assessment of a mobile game ("MobileKids Monster Manor") to promote physical activity among children. Game Health J 2015;4(2):149—58.

[53] Edwards EA, Caton H, Lumsden J, Rivas C, Steed L, Pirunsarn Y, et al. Creating a theoretically grounded, gamified health app: lessons from developing the Cigbreak smoking cessation mobile phone game. JMIR Serious Games 2018;6(4):e10252.

[54] Cafazzo JA, Casselman M, Hamming N, Katzman DK, Palmert MR. Design of an mHealth app for the self-management of adolescent type 1 diabetes: a pilot study. J Med Internet Res 2012;14(3):e70.

[55] Korda H, Itani Z. Harnessing social media for health promotion and behavior change. Health Promot Pract 2013;14(1):15—23.

[56] Sundar SS, Marathe SS. Personalization versus customization: the importance of agency, privacy, and power usage. Hum Commun Res 2010;36(3):298—322.

[57] Pichon A, Schiffer K, Horan E, Bakken S, Mamykina L, Elhadad N. Divided we stand: The collaborative work of patients and providers in a chronic enigmatic chronic disease. ACM 2020 [under anonymous review].

[58] Miller S, Ainsworth B, Yardley L, Milton A, Weal M, Smith P, et al. A framework for Analyzing and measuring usage and engagement data (AMUsED) in digital interventions: viewpoint. J Med Internet Res 2019;21(2):e10966. https://doi.org/10.2196/10966.

[59] Crutzen R, Roosjen JL, Poelman J. Using Google analytics as a process evaluation method for Internet-delivered interventions: an example on sexual health. Health Promot Int 2013;28(1):36—42. https://doi.org/10.1093/heapro/das008. PubMed PMID: 22377974.Epub 2012/03/02.

[60] Park YR, Lee Y, Kim JY, Kim J, Kim HR, Kim Y-H, et al. Managing patient-generated health data through mobile personal health records: analysis of usage data. JMIR mHealth & uHealth 2018;6(4):e89.

[61] Cheung K, Ling W, Karr CJ, Weingardt K, Schueller SM, Mohr DC. Evaluation of a recommender app for apps for the treatment of depression and anxiety: an analysis of longitudinal user engagement. J Am Med Inf Assoc 2018;25(8):955—62.

[62] Taki S, Lymer S, Russell CG, Campbell K, Laws R, Ong K-L, et al. Assessing user engagement of an mHealth intervention: development and implementation of the growing healthy app engagement index. JMIR mHealth & uHealth 2017;5(6). e89.

[63] Couper MP, Alexander GL, Maddy N, Zhang N, Nowak MA, McClure JB, et al. Engagement and retention: measuring breadth and depth of participant use of an online intervention. J Med Internet Res 2010;12(4):e52.

[64] Lin JC-C. Online stickiness: its antecedents and effect on purchasing intention. Behav Inf Technol 2007;26(6):507–16.

[65] Estrin D, Sim I. Open mHealth architecture: an engine for health care innovation. Science 2010;330(6005):759–60.

[66] Symul L, Wac K, Hillard P, Salathé M. Assessment of menstrual health status and evolution through mobile apps for fertility awareness. NPJ Digit Med 2019;2(1):64. https://doi.org/10.1038/s41746-019-0139-4.

[67] Li K, Urteaga I, Wiggins CH, Druet A, Shea A, Vitzthum VJ, et al. Characterizing physiological and symptomatic variation in menstrual cycles using self-tracked mobile health data. NPJ Digit Med 2020;3(1):1–13.

[68] Zhan A, Mohan S, Tarolli C, Schneider RB, Adams JL, Sharma S, et al. Using smartphones and machine learning to quantify Parkinson disease severity: the mobile Parkinson disease score. JAMA Neurol 2018;75(7):876–80. https://doi.org/10.1001/jamaneurol.2018.0809.

[69] Althoff T, Sosič R, Hicks JL, King AC, Delp SL, Leskovec J. Large-scale physical activity data reveal worldwide activity inequality. Nature 2017;547(7663):336–9. https://doi.org/10.1038/nature23018. PubMed PMID: 28693034.Epub 2017/07/10.

[70] Urteaga I, McKillop M, Elhadad N. Learning endometriosis phenotypes from patient-generated data. NPJ Digit Med 2020;3(88). https://doi.org/10.1038/s41746-020-0292-9.

[71] Blei DM. Probabilistic topic models. Commun ACM 2012;55(4):77–84. https://doi.org/10.1145/2133806.2133826.

[72] Blei DM, Ng AY, Jordan MI. Latent dirichlet allocation. J Mach Learn Res 2003;3:993–1022. Jan.

[73] Denny E, H Mann MC. A clinical overview of endometriosis: a misunderstood disease. Br J Nurs 2007;16(18):1112–6.

[74] Price RK, North CS, Wessely S, Fraser VJ. Estimating the prevalence of chronic fatigue syndrome and associated symptoms in the community. Publ Health Rep 1992;107(5):514.

[75] Wood B, Van Der Mei I, Ponsonby A-L, Pittas F, Quinn S, Dwyer T, et al. Prevalence and concurrence of anxiety, depression and fatigue over time in multiple sclerosis. Mult Scler J 2013;19(2):217–24.

[76] Louati K, Berenbaum F. Fatigue in chronic inflammation-a link to pain pathways. Arthritis Res & Ther 2015;17(1):254.

[77] Rabbi M, Pfammatter A, Zhang M, Spring B, Choudhury T. Automated personalized feedback for physical activity and dietary behavior change with mobile phones: a randomized controlled trial on adults. JMIR mHealth & uHealth 2015;3(2):e42.

[78] Ambeba EJ, Ye L, Sereika SM, Styn MA, Acharya SD, Sevick MA, et al. The use of mHealth to deliver tailored messages reduces reported energy and fat intake. J Cardiovasc Nurs 2015;30(1):35–43. https://doi.org/10.1097/JCN.0000000000000120. PubMed PMID: 24434827.

[79] Schembre SM, Liao Y, Robertson MC, Dunton GF, Kerr J, Haffey ME, et al. Just-in-Time feedback in diet and physical activity interventions: systematic review and practical design framework. J Med Internet Res 2018;20(3):e106. https://doi.org/10.2196/jmir.8701.

[80] Holmen H, Wahl AK, Cvancarova Småstuen M, Ribu L. Tailored communication within mobile apps for diabetes self-management: a systematic review. J Med Internet Res 2017;19(6):e227. https://doi.org/10.2196/jmir.7045.
[81] Druce KL, Dixon WG, McBeth J. Maximizing engagement in mobile health studies: lessons learned and future directions. Rheum Dis Clin 2019;45(2):159–72. https://doi.org/10.1016/j.rdc.2019.01.004.

CHAPTER 6

Mobile health apps: the quest from laboratory to the market

Juan José Lull, PhD, Antonio Martínez-Millana, PhD, Vicente Traver, PhD

Instituto Universitario de Investigación de Aplicaciones de las Tecnologías de la, Información y de las Comunicaciones Avanzadas (ITACA), Universitat Politècnica de València, Spain

Contents

Digital Health
ISBN 978-0-12-820077-3
https://doi.org/10.1016/B978-0-12-820077-3.00006-7

1. Introduction

In the framework of the MyCyFAPP Project [1] that was funded by the European Union, a process was followed that allowed us to determine a methodology for the development of a mobile app, from broad requirements to the distribution of the intellectual property (IP).

We decided to follow a repeatable process that would lead to decisions for each stage of the product. The central element was an app with the aim of helping in the self-management in patients with cystic fibrosis (CF).

The need was clear: CF is a genetic, progressive disease that affects more than 70,000 people worldwide, according to the Cystic Fibrosis Foundation [2]. The disease causes lung infections and affects the digestive system. The aim of the MyCyFAPP project was to develop an app that helped patients and caregivers with one important problem, enzyme regeneration: The pancreas does not release digestive enzymes, and they have to be administered with food intake. Basically, the app would help the patients adjust their enzyme dosage in their pancreatic enzyme replacement therapy (PERT) [3].

2. Business model

Finding and selecting business models in the healthcare sector is a complex process because it usually involves a combination of public customers with concrete and disparate requirements. Specifically, providers tend to be private entities that deliver highly tailored products to the particular constraints of the healthcare provider. Therefore, the best strategy is to select a methodology that allows entrepreneurs and innovators to implement an incremental process by which the hypotheses are refined and evolved. For that reason, we selected, as a reference methodological framework, the Business Model Generation Canvas (BMGC [4]), combined with the "Disciplined entrepreneurships: The 24 steps toward a successful start-up" (DE24 [5]).

- BMGC: This methodology simplifies the identification of the nine most relevant aspects of a certain business model: value proposition; customer segments; revenue streams; distribution channels; key activities; key partners; cost structure; key customer relationships; and competitors.
- DE24: It describes a methodology by which the business model is refined and extended into a strategy that allows market entry in a successful way, having in mind that the assets intending to be exploited are new in the addressed market.

As a first step to apply these frameworks, we needed to learn about the market dynamics, perspectives, needs, concerns, available tools, support for patients and professionals, etc. Gathering all this information would allow the business models to be better adapted to the real situation and opportunities of the customer segments proposed.

The process will be explained in the next section.

3. Context: business models

In order to identify our business model, we first needed to perform field interviews. The most relevant requirements were collected from patients, families, and professionals. Afterward, an analysis was performed for CF coverage and healthcare system. The main focus was on European systems implemented to support patients and family. Other relevant regions like Australia, Canada, the United States, and India were also considered.

Available apps with an aim at CF and those that could be used by the community around CF were also analyzed.

Three potential business models were identified:

1. Hospital-driven CF service,
2. Self-management, and
3. Game/training.

3.1 Hospital driven CF service

The asset for this business model was the MyCyFAPP system including all the software artifacts (apps, games, and PWT), enriched food databases (MyFoodREC), and algorithms (personalized enzyme replacement dosage per meal) together with the adapted care procedures and the professional service that supports the whole intervention program with patients and/or their families. The service, patients, and families would thus be the users for this asset.

The clients would be the CF units in public and private hospitals.

3.2 Self-management

In this business model, it is assumed that there is no CF unit service behind the App, and that the end users consider the App (and games) as an everyday tool that help them track symptoms, register food intakes, access individualized recommendations of enzymes per meal, and manage their own adapted recipes. Opposite to the previous business model, there is no healthcare professional looking at the information uploaded by the patient or family members.

Clients in this case could either be patients and families or sponsors.

3.3 Game/training

In this case, the assets to be exploited are games for children's education and training, as well as the guidelines to be followed in MyCyFAPP app.

4. From business models to products

The definition and development of the MyCyFAPP solution is a complex task, as it needs to go through data collection, software development iterations, piloting activities, and validation in clinical trials in multiple clinical centers.

The exploitation strategy focuses on guiding the formulation of the different 80 business models in which the solution can be commercialized.

Nevertheless, the MyCyFAPP solution is composed by six different modules, which can also be exploited on an individual basis.

This section contains a description of these individual products that show their main characteristics. On a business basis, each of the individual exploitable products are described, taking into account the lean canvas main features and avoiding others, such as the cost structure, revenue streams, channels, and customer segments.

Each product has its own description with the following information:
1. Who the users and clients would be,
2. The main features of the product and possible competitors,
3. The SWOT Analysis (Strengths, Weaknesses, Opportunities, Threats),
4. The differential advantage (what makes the product different to potential competitors),
5. The unique value proposition (what it is and what it is good for).

The process that was followed in order to move each business model into exploitable results appears in Fig. 6.1.

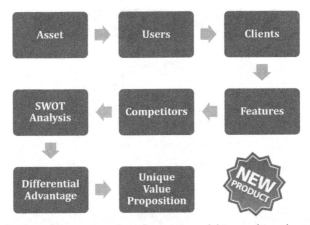

Figure 6.1 Outline of the process from business model to reaching the new product.

The exploitable products that were defined through the study of the different aspects were

1. MyFoodREC: Online database system to collect electronic case reports on food intake.
2. Nutritional Recommendations Handbook (NRH).
3. Personalized Optimal Enzyme Dose Calculator.
4. MyCyFAPP: Mobile application.
5. MyCyFAPP: Gaming application.
6. MyCyFAPP: Health Professional Web Tool (PWT).

At the end of the chapter, two use cases are presented with information about two of the products, the mobile app and the NRH.

5. Product dependencies

The MyCyFAPP system should be understood as a comprehensive holistic solution to empower patients, promote self-management, and provide clinicians with tools to perform a continuous remote control of their patients, as seen in Fig. 6.5 and Fig. 6.2.

Even though the analysis of the exploitation products was performed on a one-by-one basis, there are critical dependencies among these products that should be clarified. Fig. 6.3 shows a schema of the identified interdependencies. The approach to effectively manage dependencies is achieved by defining a methodology to deal with dependencies that works across all products. Traditionally these dependencies have been described as appearing between the following product attributes: **function, properties,** and **structure**.

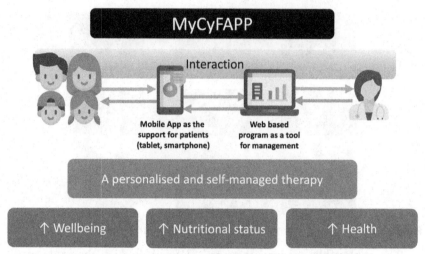

Figure 6.2 Interactions around the MyCyfApp mobile app.

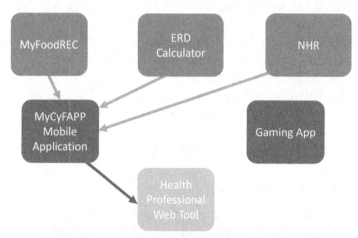

Figure 6.3 MyCyFAPP individual products dependencies.

The MyCyfAPP sum of all individual products should be seen as a whole and the marketing strategy and IP rights (IPR) had to be approached having this in mind. Afterward, we studied the business scenarios that could lead to the best approach for disseminating the products. Finally, the distribution strategies of the software were analyzed.

6. Marketing strategy

Marketing activities were started as soon as the products were defined with the proposed methodology.

The ultimate goal of sharing the vision from the beginning to the entire set of primary, secondary, and tertiary stakeholders could only be reached by adopting a multidisciplinary framework, in which the marketing actions would be triggered by all consortium partners in different settings and to different audiences.

The first stage of the project development was focused on the identification of the stakeholders and end users, and thereafter, using a semistructured interview methodology to perform an elicitation of their needs and requirements to define the project use cases, and moreover, mock-up each of the products that would enable a user interaction. The first version of the exploitation plan was used to classify and define the products from a business perspective, and in the following months, it helped redefine and tune the applications, mainly focusing on leading them to their commercial exploitation as research, clinical, and personal products.

Once the project's main concepts and products had been defined, using the functional prototypes and working prototypes, the strategy focused on assessing its commercial viability by validating in specific focus groups, questionnaires, and user tests with a broader end users and clients audience from all the different perspectives (clinical, usability, technical performance, and business added value).

Finally, project early adopters (hospitals, healthcare agencies, and CF patients associations) collaborated to test and validate the deployment of the entire MyCyFAPP system in real settings.

7. Intellectual property rights

Aside from the marketing strategy, the product needed a distribution of the IPRs, since there were six individual products and 10 partners that had worked in the different stages with the development from the inception (Table 6.1).

The project, being funded by the European Union with a Grant Agreement, had a default IPR protection during the project time span. Thus, the different partners needed to devise a protection strategy for the products.

The different partners were surveyed asking for their participation in the different parts of the project. This could also be somehow contrasted by the hours dedicated to the project that every partner had registered, although both things do not exactly correlate.

Table 6.1 Project integrating partners.

Partner	Partner short name
Instituto ITACA, Universitat Politècnica de València, Valencia, Spain	ITACA UPV
Instituto de Investigación Sanitaria del Hospital Universitario y Politècnico La Fe, Valencia, Spain	HULAFE
Youse GmbH, München, Germany	YOUSE
SINTEF, Trondheim, Norway	SINTEF
Department of Architecture and Technology of Computers, University of Seville, Seville, Spain	
Pediatric Pulmonology and Cystic Fibrosis Unit, Department of Pediatrics, University Hospitals Leuven, Leuven, Belgium	KUL
CF Center, Università degli Studi di Milano, Fondazione IRCCS Ca' Granda, Ospedale Maggiore Policlinico, Via Commenda 9, Milan, Italy	USM
Hospital Universitario Ramón y Cajal, Carretera Colmenar Viejo km 9, Madrid, Spain	SERMAS
ERASMUS MC, Dr. Molewaterplein 40, 3015 GD Rotterdam, the Netherlands	ERASMUS MC
Association for Research and Development of the Faculty of Medicine, Lisbon School of Medicine—Universidade de Lisboa, Lisbon, Portugal	AIDFM
CF Europe	CFE
SINTEF, Forskningsveien 1a, Pb. 124 Blindern, Oslo, 0314, Norway	SINTEF
Imaginary, Piazza Caiazzo, 3 20124 Milano, Italy	IMAGINARY

For every product, the participation was reported in the following categories:
• Concept definition
• Concept development
• Content development
• GUI/elements
• Implementation (software)
• Validation

As an example, for the MyCyFAPP mobile app, the roles were assigned to the different partners, as seen on Table 6.2.

8. Business scenarios

Once the products had been seen as viable in the market and the results had been protected, one or more commercialization strategies had to be followed.

Table 6.2 IP definition for the MyCyFAPP.

Result	Type (SW-REPORT)	Ownership
General concept of app (concept and overall features)	Software Requirements Specification	HULAFE
User requirements	Software Requirements Specification	YOU SINTEF TSB
Mock-ups	Design of the application (Balsamiq, videos)	SINTEF
Architecture	Design of the app	SINTEF TSB
Graphical elements	Images	HULAFE
Configure and set preferences	Software	SINTEF UPV ITACA
Summary view	Software	SINTEF
Food diary layer	Software	UPV ITACA
Manage enzyme intake Layer	Software	UPV ITACA

Many apps around CF have been developed, as we will see in the sections about Competitors, specifically in 10.1.4 and 10.2.4, but the review we conducted shows that they have limitations [6]: none of them were found to be appealing to the user; none of them supported PERT; and the user needs are only partially matched by the existing apps.

Furthermore, most of the apps that were reviewed have disappeared because of lack of a market structure that led developers, clinicians, users, customers, etc., to an understanding and collaboration.

In order to have a successful self-management treatment app that users could effectively use, possible business scenarios had to be explored.

We found three feasible scenarios that would effectively enable self-management of PERT. They differ in who is in charge, what channels exist from the exploitation group to the users and clients, etc.

A basic summary of the scenarios is shown in Fig. 6.4. As can be seen, the IPR holder is the initial node and that reflects us, the ones in charge of the project. Afterward, another company takes charge of the product (scenario 2, Party B in the figure), another party with access to needed infrastructure takes the lead of the distribution (Party A in the figure), or an exploitation team is created inside the project and the distribution and implantation would be based on a sponsor institution (Party C). These are the scenarios, and they are explained at the following sections.

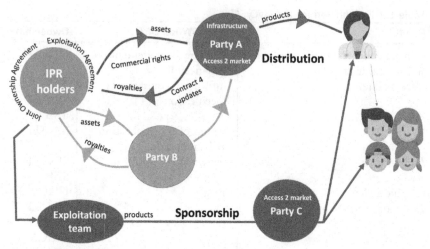

Figure 6.4 MyCyFAPP Business Scenarios. Blue color (dark gray color in printed version) for Scenario I; green color (light gray color in printed version) for Scenario 2; and orange color (gray color in printed version) for Scenario 3.

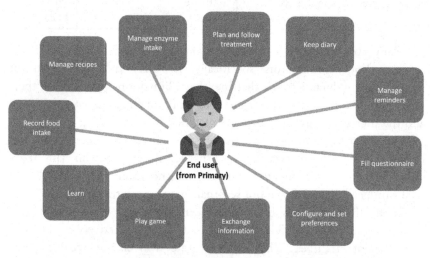

Figure 6.5 Features in the mobile app to satisfy user needs.

8.1 Distribution through CF units and hospitals: Scenario 1

In this business scenario, a company with all the regulatory and techno-logical requirements should distribute the MyCyFAPP solutions to users. This company should be an international company with market presence and access to market channels.

The commercial rights of the MyCyFAPP solutions (all the products and assets) are licensed under a nonexclusive contract in which IP owners allow the company to distribute the solution to users through CF units and hospitals.

It is mandatory that the company has the regulatory permissions to distribute medical technology (Medical Device Class 1) and that they have the infrastructure to host the back-end services of the software components (self-management app, web tool, and game).

This business scenario includes other types of contracts for product maintenance, updates, and the improvement to the IP owners.

In this way, the distributor has a set of licenses (alphanumeric codes), which are distributed to the doctors, together with the links to access the PWT. For example, the children with CF (or their relatives) download the App or the Game from an official app market and install it. Both doctor and patient are trained using the materials and multimedia generated during the project. Sensitive data are securely stored under the responsibility of the distributor, who has to grant the accomplishment of General Data Protection Regulation and CE mark of a Medical Device Class I requirements.

8.2 Business to bussiness: Scenario 2

In this scenario, the commercial rights and the ownership of the products are transferred to a third party that is interested on the commercial exploitation of one (or more) MyCyFAPP products. This third party may not distribute the products to the users but to clients. These can be pharma companies, clinical institutions, and research groups.

The transfer is bound to a contract in which IP owners would receive a year royalty or an economic compensation.

This business scenario is not included in the MyCyFAPP Business Vision, as it does not grant easy access to the self-management App and the Game to children with PCF, but could be interesting in terms of individual products exploitation, such as nutritional databases, parts of the software, and the algorithm.

8.3 Sponsorship: Scenario 3

In this scenario, part of the MyCyFAPP Project consortium (an exploitation team) invests in creating a commercial product based on the outcomes of the research project. This scenario is similar to Scenario 1, but in this case, the distribution company does not have the required technological resources to host the software solutions and the regulatory level to certify the App as a medical device.

In contrast with Scenario 1, the involvement of a third party is needed to sponsor the costs of maintaining, updating, and customizing the solution, but there are no agreements to license the IP and commercial rights, as these would be kept by the original owners, the IPR holders.

The distribution company should have the regulatory clearance to distribute medical technology and access to the market channels (commercial relationship with CF Units or Hospitals).

9. Distribution: exploitation models

There are several models for the exploitation of a mobile app. Basically, focusing on the Apple App Store and Google Play Store, we could divide the revenue sources as

1. Payment per app install: The app has a price and there is a single payment. A big drawback for this revenue model is that there is a price barrier and the number of users who download the app tends to be low.
2. Freemium model: The app install is free. However, key (premium) content is accessible through in-app payment.
3. Virtual assets: Different parts of the app could be diversified in sections that needed a payment for each one.
4. In-app subscription: A payment per year or per month would enable access to the self-management PERT app. The mechanism for subscription would be the one already existing in the Apple and Google stores.
5. Subscription through the PWT: Access to the platform would be enabled through the PWT (e.g., an access key would be bought either per hospital, per patient, etc.). This key would be used inside the app in order to access most of the contents, including the PERT management.
6. App sale or sponsorship: An institution could be interested in purchasing the app or sponsoring it. Depending on whether it was one or the other, a revenue model should be established.
7. Consultancy to hospitals: The app would be free but the installation of the accompanying PWT and server maintenance, etc., would be carried out by the MyCyFAPP Consortium.
8. Customized app for each hospital: The app would be encapsulated into a software development kit so each medical center could have their own app. That would help each center have a proprietary app with their own database and branding layout.
9. Advertisement platforms: Advertisement could be established in the possible platforms, targeting the population with specific rules by defining rules about the ads, such as population age.

Some other typical revenue strategy possibilities could be defined, but most would be combinations of the previous ones.

10. Product use cases

In the section of business model to products, we saw the methodology we applied. Two products have been chosen and are described in the following subsections:

10.1 MyFoodREC, nutritional database

10.1.1 Users
Patients, doctors, and nutritionists.

10.1.2 Clients
Database companies, software development companies, healthcare service providers (public/private), and pharmaceutical companies.

10.1.3 Features
Enhancement on the food items and nutritional composition of two licensed databases (EuroFIR [7]/Nubel [8]) by refining and completing missing indicators and items.

Food items are classified according to the diet characteristics of five European countries (Spain, Portugal, the Netherlands, Belgium, and Italy).

The food composition has the following items: energy, proteins, carbohydrates (CH_2O), starch, sugar, fat, monounsaturated fatty acids (MUFAs), polyunsaturated fatty acids (PUFAs), saturated fatty acids (SFA), total fiber, iron (Fe), calcium (Ca), and sodium (Na).

10.1.4 Competitors
- **Dietowin:** A computer nutritional tool created for the realization of personalized diets tailored to any patient, in a few seconds, with all the scientific guarantees (www.dietowin.es).
- **Easydiet:** It includes nutritional composition databases, personalized nutritional reports, etc. In collaborations with the Spanish Association of Nutritionists/Dietitians. Sponsored by Biocentury SL: (www.easydiet.es).
- **Nutrimind:** It includes nutritional composition databases, personalized nutritional reports, etc. Nice reports with graphics. It allows for the creation of dishes, in a similar way as MyFoodCALC (www.nutrimind.net).

10.1.5 SWOT analysis
See the Table 6.3.

Table 6.3 SWOT analysis of MyFoodREC.

Strengths	Weaknesses
- Unique database specific for CF.	- Lack of involvement from pharmaceutical companies, public policymakers and the insurers' domain.
- Endorsed by clinical and scientific specialists specific for regional families of dishes.	- It lacks some of EU countries' regional dishes.
- Complete recordings on several composition items.	
Opportunities	**Threats**
- Enhancement of existing databases.	- Based on licensed databases, IPRs only assignable for the created entities.
- Complementing existing systems and nutritional databases for CF and other diseases.	- The availability or adoption of certain medications by public funding insurers could decrease the need of having a nutrition-based app for self-management.
- Open to new updates.	

10.1.6 Differential advantage

- Nutritional facts are adapted to CF patients' specific needs and preferred dishes.
- Food products are named and labeled to be easily found and identified.
- MyFoodREC includes, for the first time, nutritional facts for all the items in the database, filling the gaps of the licensed former nutritional databases (specifically, EuroFIR and Nubel were completed with nutrient data for sugar, SFAs, MUFAs, PUFAs, fiber, calcium, iron, and sodium) [9].
- Allows for complete nutritional intake studies.
- Allows for long-term nutritional intake studies.
- Allows for real-time calculation of the nutritional evaluation report.

10.1.7 Unique value proposition

A comprehensive nutritional components database specifically designed for CF and country-specific dishes.

10.2 MyCyFAPP: mobile application

10.2.1 Users

Patients and doctors.

10.2.2 Clients
Software development companies, healthcare service providers (public/private), pharmaceutical companies, and patients associations.

10.2.3 Features
MyCyFAPP mobile application has five main modules:

Diet assessment: MyCyFAPP is equipped with the MyFoodREC database and with a calculation algorithm that allows for a complete analysis of the nutritional composition for each meal.

Adjust enzymes dosage: The integrated enzyme dosage calculator model acts like an enzyme dosage predictor for each meal. Patients introduce in MyCyFAPP the food products or dishes they are going to eat, and MyCyFAPP indicates, in real time, the optimal dosage for the particular meal.

MyResults: Patients are able to watch, at any time, their records concerning nutritional composition of their food records, the actions they have performed so they can learn how to solve complications, the amount of enzymes predicted for previous meals, etc.

FAQs and feedback: Patients can check the NHR and frequently asked questions. If a question does not appear in the list, they can ask about their concern through a message to a health professional, who will periodically check and answer questions from the patients. The documentation also contains general tips and recommendations for an adequate diet for pediatric CF patients and specific solutions for specific complications too.

Quality of life assessment: Assessed by means of a quality of life questionnaire and also by questions that pop up in the software such as "Any abdominal pain?", "Did you enjoy your lunch today?".

10.2.4 Competitors
- **Portable Genomics:** This provides individuals with control of their personal health data in order to facilitate sharing the data with healthcare providers and life science organizations leading to improved healthcare and smarter health discovery. www.portablegenomics.com
- **Genia:** It is an app for youngsters living with a long-term illness. It helps you keep track of your health and communicate better with your family, healthcare team, and peers (other people with CF). Genia was founded by Andreas Hager whose daughter, Sonia, was diagnosed with CF when she was 5 months old (www.genia.se).

- **Cystic Fibrosis: a pocket guide.** An interactive and educational app developed by nurses in Dundee. The app has been developed with partial funding via educational grants from Roche Products Ltd. and Forest Laboratories UK Ltd (Link to the pocket guide for iBooks).
- **CF GeneE:** Educational mobile application for healthcare professionals. It allows the user find information about common mutations in the cystic fibrosis transmembrane conductance regulator gene that leads to CF disease (www.cfgenee.com).
- **CFBuzz:** It is an app built for adults with CF, and it helps assist in CF self-management. It creates a personal medical ID and helps establish goals and health plans: FEV1, HbA1C, weight, BMI, vitamins, etc. It tracks scripts and appointments (www.cfbuzz.org/app).
- **My Fight Against Cystic Fibrosis:** General education of the disease and population statistics. Latest news about the disease. Common treatment providers' list and contact information. Medication tracking log lets you store daily medications (My Fight Against CF App)
- **MyTherapy:** It is a pill reminder app for all medications. It has a logbook for skipped and confirmed intakes. It has support for a wide range of dosing schemes within medication reminders. Tablets, dose, measurements, and activities can be tracked for all conditions (diabetes, rheumatoid, arthritis, anxiety, depression, hypertension, multiple sclerosis). Measurements include weight, blood pressure, and blood sugar levels. A journal may be printed and shared with the doctor (www.mytherapyapp.com).
- **CF medication App:** The app calculates CREON dosage based on the amount of fat in product and sets a dose of CREON per gram of fat. It is only available in Dutch and iOS (link removed from the App Store).
  - Squeezy **CF:** The app helps men and women with CF remember to do their pelvic floor exercises and to do them in the right way. The app is endorsed by the CF Trust (https://www.nhs.uk/apps-library/squeezy-cf/).
  - **Fast Tract Diet App for Gut Health:** It "helps you identify gut friendly foods versus hard-to-digest foods, track and chart your meals and symptoms, create shopping lists and quickly look up symptom potential for specific foods & drinks at your favorite market on the fly." It is based on a flexible "FP" point system adaptable to a variety of dietary preferences. The FP score is calculated based on sugar alcohols, total carbohydrate, and dietary fiber. Thus, the app does not take lipid ingestion into account. The app claims to be friendly for CF

patients, but it does not help in the nutritional aspects that are relevant to digestion problems inherent to CF (https://digestivehealthinstitute. org/2015/12/17/fast-tract-diet-app/).

- **Perx—feel rewarded:** This app is a gamification software that turns the chronic patients' chores into challenges that earn them rewards. It is connected to a back end by Perx Health. There are Australian chronic patients with different conditions that use the app, and one subgroup is the one composed by patients from the Cystic Fibrosis Australia Association Federation. The app helps the user with alarms and tasks that the user needs to introduce (https://www.cysticfibrosis. org.au/perx).

- **Cystic Fibrosis by AZoMedical:** This is an informational/educational app that was apparently discontinued in 2015 (https://apps. apple.com/us/app/cystic-fibrosis-by-azomedical/id718172748).

A comprehensive study about possible app competitors has been recently conducted [6].

10.2.5 SWOT analysis
See the Table 6.4.

Table 6.4 SWOT: MyCyFAPP mobile application.

Strengths	Weaknesses
- The functionality of the app is unique, letting the patients self-manage the enzyme dosage for each particular meal.	- It depends on user motivation and willingness to manage CF.
- Integration of users into the development process.	- Lack of involvement from the pharma companies, public policymakers, and insurers domain.
- Based on Behavior Change Techniques to promote patient self-management and awareness.	- Lack of interoperability standards.
- Endorsed by users in a validation pilot based on the IEEE 1471 software architecture.	- Lack of reimbursement policies and financial incentives for healthcare providers, since current budget allocation in the public sector is still focused on the number of patients and not in the organizational aspects or in the efficiency of the processes.

(Continued)

Table 6.4 SWOT: MyCyFAPP mobile application.—cont'd

- Developed upon deep use Case definition. - An agile software development is adopted aiming at developing and testing incrementally the apps.	
Opportunities	**Threats**
- It can complement existing systems for CF management and other diseases.	- Competitors.
	- Lack of cohesion at healthcare system across EU countries.
- Easy to integrate with databases, Personalized Enzyme Dose Calculator, NHR, and web application.	- Lack of clarity on mHealth certification: healthcare professionals do not prescribe technology because of the lack of a clear certification regulation that promotes the use of safe and reliable mobile applications. - It depends on users' motivation and willingness to manage CF.
- Better nutritional management and adjusted enzymes dosage is linked to less infections and better lung function.	- The availability or adoption of certain medications by public funding insurers could decrease the need of having a nutrition-based app for self-management.
- Easy edition as stand-alone. - Many apps focus on scientific information and contents for education and are promoted from reference organizations like EU CF Society. In those cases, the content is not for patients but for researchers. - There are several apps, but none of them have the ambition and focus of MyCyFAPP - Patients are looking forward to being supported.	

10.2.6 Differential advantage

- Context aware development to assure that the developed innovation is based on real CF user needs.

Designed for specific target groups: Children 4–12 years old who will not use the app by themselves but by their parents; teenagers 13–16 years old; adult patients; and supervisors (parents and doctors).

- Focused on replacing enzymes, one of the most recurrent discussion topic among CF patients.
- Compliant with Helsinki Declaration with regards to patient safety and privacy.

10.2.7 Unique value proposition

A mobile application tailored for each of the specific needs of CF patients' age segments, which integrates MyFoodREC, NHR, and Personalized Optimal Enzyme Dose Calculator to empower patients, and at the same time, it supports the self-management of the disease.

11. Conclusion

The process of defining and creating a mobile health product was defined. The unique app-centered solution autonomously prescribes the optimal supplemental enzymatic dosage for CF patients, based on their meals. They can introduce their meal data and the enzymatic supplement is automatically offered to them.

The method, from inception till dissemination, has been described. We tried to establish a method that could be helpful for the MyCyFAPP project. In addition, the process should have a scope wide enough so that other entrepreneurial projects could follow it. Some methods may be followed by others in exactly the same way and same order as they were conducted in the project, while others will need adaptations. The different steps, along with an actual implementation, have been presented. This methodology could be followed, especially if many partners are concerned in the development of products around an app.

References

[1] The MyCyFAPP Consortium, MyCyFAPP Project, URL https://www.mycyfapp.eu/index.php/en/.
[2] Cystic Fibrosis Foundation, About Cystic Fibrosis, URL https://www.cff.org/What-is-CF/About-Cystic-Fibrosis/.
[3] Sikkens EC, Cahen DL, Kuipers EJ, Bruno MJ. Pancreatic enzyme replacement therapy in chronic pancreatitis. Best Pract Res Clin Gastroenterol 2010;24(3):337−47. https://doi.org/10.1016/j.bpg.2010.03.006.
[4] A. Osterwalder, Y. Pigneur, T. Clark, A. Smith, Business model generation: a handbook for visionaries, game changers, and challengers, Wiley.
[5] Aulet B. Disciplined entrepreneurship: 24 steps to a successful startup. Wiley; 2013.
[6] Martinez-Millana A, Zettl A, Floch J, Calvo-Lerma J, Sevillano JL, Ribes-Koninckx C, Traver V. The potential of self-management mHealth for pediatric cystic fibrosis:

mixed-methods study for health care and app assessment. JMIR mHealth uHealth 2019;7(4):e13362. https://doi.org/10.2196/13362. URL, http://www.ncbi.nlm.nih.gov/pubmed/30998222http://www.pubmedcentral.nih.gov/articlerender.fcgi?artid=PMC 6495294.

[7] Finglas PM, Berry R, Astley S. Assessing and improving the quality of food composition databases for nutrition and health applications in Europe: the contribution of EuroFIR. Adv Nutr 2014;5(5):608S–14S. https://doi.org/10.3945/an.113.005470. URL, https://academic.oup.com/advances/article/5/5/608S/4565780.

[8] Seeuws C. Belgian Branded Food Products Database: inform consumers on a healthy lifestyle in a public-private partnership. J Food Compos Anal 2017;64:39–42. https://doi.org/10.1016/j.jfca.2017.07.008.

[9] Calvo-Lerma J, Hulst J, Boon M, Martins T, Ruperto M, Colombo C, Fornés-Ferrer V, Woodcock S, Claes I, Asseiceira I, Garriga M, Bulfamante A, Masip E, Walet S, Crespo P, Valmarana L, Mart ınez-Barona S, Pereira L, de Boeck K, Ribes-Koninckx C. The relative contribution of food groups to macronutrient intake in children with CysticFibrosis: a European multicenter assessment. J Acad of, Nutr Dietet 2019;119(8):1305–19. https://doi.org/10.1016/j.jand.2019.01.003.27.

CHAPTER 7

mHealth in public health sector: challenges and opportunities in low- and middle-income countries: a case study of Sri Lanka

Pamod Amarakoon, MBBS [1], Roshan Hewapathirana, MBBS, MSc, PhD [2], Vajira H.W. Dissanayake, MBBS, PhD [2]

[1]Postgraduate Institute of Medicine, University of Colombo, Colombo, Sri Lanka; [2]Faculty of Medicine, University of Colombo, Colombo, Sri Lanka

Contents

Digital Health
ISBN 978-0-12-820077-3
https://doi.org/10.1016/B978-0-12-820077-3.00007-9

1. Introduction

Technology has brought about revolutionary reforms to the ordinary practices of all domains. Healthcare is no exception. This has been facilitated by the accessibility of technology to users in last decade. The computer is the primary mode of interaction with technology for humans. Gradual decrease in size of the device has transformed computers that are mobile and can fit into the size of a pocket. In modern world, smart mobile device is the primary point of contact of technology for individuals which has escalated ease of access and the portability. The letter "m" is associated with use of mobile technology for existing domains. mHealth is a term that is used increasingly coupled with technology-associated healthcare in modern terms. mHealth or mobile health could be defined as medical or public health practice supported by mobile devices [1]. The main advantage of using mobile technology is to speed up the communication process in healthcare delivery and to expand the reach of healthcare to areas which were conventionally not possible due to limited access and unavailability of resources. The type of device used has changed over time from personal digital assistants to mobile phones, wearables, and other sophisticated mobile devices. The initial era of mHealth used short messaging service (SMS) and USSD codes for transmitting information between the mobile device and the central information hub. Wireless networking and fast mobile networks are commonly used in recent times to communicate with the devices.

With speed of access to healthcare being a major factor contributing to the establishment of mHealth technology, it is widely used in curative healthcare domain for patient monitoring and provision of telehealth, the latter term used to refer the use of telecommunication and virtual technology to deliver healthcare outside the health facilities [2]. This is used effectively in detecting acute medical conditions such as strokes in remote areas [3]. This has to a certain extent shadowed the application of mHealth to a less acute, yet more important in community perspective; the domain of public health. Public health is "the art and science of preventing disease, prolonging life and promoting health through the organized efforts of society" as defined by Acheson in 1988 [4]. There are many aspects of mHealth use in public health domain which include public health surveillance, monitoring and evaluation of public health programmes, and coordinated organizational activities and online communications [5]. Increased penetration of mobile technology in the developing world in the last decade has created an opportunity to use this platform to cater public health requirements.

As of 2019, the World Bank defines countries with a gross national income (GNI) of $1025 per capita or less and low-income countries and countries with a GNI of $1026 to $3995 per capita as lower middle–income countries [6]. Countries that fall into low-income and lower middle–income countries are together classified as LMICs [7].LMICs encounter many barriers in enhancing their health services mainly due to limitations in financial, human resources, and materials [8]. Many LMICs have attempted using technology to overcome barriers in delivery and monitoring of health services but have found it challenging to scale up and sustain the innovations [9]. However, there were more than 400 mHealth projects operating in Africa alone by year 2015 [10]. If these health innovations were to be successful, it would certainly contribute to enhance the quality of healthcare in LMICs in general. Out of numerous contributing factors regulatory, technological, and user-related concerns have been identified as inhibiting scaling up of mHealth-related innovations in LMICs [11]. Since technology becomes a major determining factor in mHealth implementations, it is worthwhile to understand the common technological barriers that I encountered in LMICs in general. Inadequate cellular mobile infrastructure, prohibitive costs, and unreliable technology have been identified as key constraints of applying technology [11]. Strength in recent years has experienced significant boost in technological adoption and expansion especially in the post-conflict era. Smartphone adoption was 0.3 million in the year 2009, while it has exponentially increased to 11.8 billion by year 2017, 8 years into the post-conflict period. Penetration of 4G technology into the cities as well as suburbs in Sri Lanka has enabled better quality connectivity to the clients. In addition, there are a number of projects mediated by the government along with the support of the private sector to expand the technology to the general public [12]. Yet a lot more to be done with regard to distribution of infrastructure across the entire to avail the technological advances to the whole population. Infrastructure is not equally distributed across the country. In addition to that improving digital literacy, addressing gender disparities in accessing digital technology, enhancing proper technical standards in web and digital applications, and increase concerned over cybersecurity have been identified as challenges with regard to technological expansion and adoption in Sri Lanka [12].

In this chapter, we will focus on use of mHealth in a setting of LMICs by using an example from Sri Lanka to address the problem of childhood malnutrition.

Sri Lanka's development of health sector has been remarkable, and sufficient proof is provided by its high-performing indicators which are sometimes better than those of developed countries and has been highlighted as a role model for the development of healthcare systems of the country [12]. Country's performance in controlling communicable diseases, in enhancing maternal and child healthcare, and its approach to preventing noncommunicable diseases has been outstanding. Yet, one area the country's health system is not satisfied is its performance in domain of nutrition. Fig. 7.1 highlights the progression of indicators related to childhood malnutrition over the years in Sri Lanka. It is quite obvious that prevalence of stunting and underweight are significantly high in Sri Lanka and the rates have remained static without much improvement over the last twodecades. This highlights the urgent need to reveal the policy and approach in addressing the issue of malnutrition.

Malnutrition can result in grave consequences to the country's economy and development. Childhood malnutrition results in a child being exposed to recurrent illnesses and the child not being able to achieve his full potential of cognitive development limiting his contribution to the development of the country in adult life. In addition, the malnourished child growing into adulthood is likely to deliver another malnourished baby. This vicious cycle is invariably affecting the country's context if a significant proportion of the population is malnourished [13]. In Sri Lankan context, poor nutrition can result in recurrent infections and exposure to disease with increased burden on healthcare and low performance in day-to-day activity, in general, can result in heavy burden on the economy [14]. It has been understood that socioeconomic status of the family

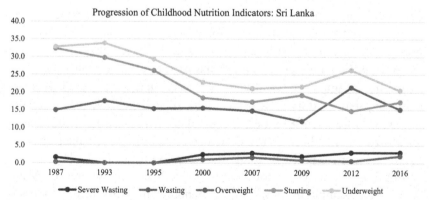

Figure 7.1 Progression of childhood nutrition indicators: Sri Lanka. *(https://data.unicef. org/topic/nutrition/malnutrition/).*

including size of the family, the size of the household, economic status of the family, and mother's education status and employment have contributed to the child's malnutrition status [14].

In spite of the seemingly nature of malnutrition as a health issue, this results from deficiencies from multiple domains in social context [15]. In Sri Lankan context, they have identified several domains such as health, agriculture, fisheries and livestock, education, and social services as the sectors that have a closer link with nutrition. Hence, deficiencies in each of this field are identified as risk factors for malnutrition. A sequential approach to addressing issue of malnutrition requires identification of children with malnutrition, risk factors contributing to malnutrition and interventions across multisector domains. Multisector nature of the problem proves that a blanket approach with a package of interventions, which has proven to be ineffective in the long run, has, in turn, necessitated interventions to be done at individual level. The whole scope of capturing data related to multisector risk factors and interventions at individual level makes identification and transfer of information between stakeholders a complicated process. Thus, execution of traditional information flow from identification of malnourished child by field-level healthcare worker to transmission of risk factor information across multisector stakeholders and final intervention takes a considerable amount of time. This results in interventions performed at individual level based on outdated information, aggravating the problem of malnutrition and creating a new set of issues at individual level which may require a different set of interventions, resulting in a vicious cycle.

The proposed approach of mHealth solution required the capture of sociodemographic, relevant nutritional and risk factor—related information at field level by the healthcare worker assisted by the field-level multisector staff using a mobile application. Once this is performed, data are made available to supervising officers of field health staff upon synchronization with central server. The data will then be checked by the supervising officer for accuracy and validity. The validated data will be made available to multisector staff of operational capacity at subdistrict level who can initiate a series of interventions based on individual risk factors (Fig. 7.2).

While the multisector risk factor assessment and interventions take place, grassroot level health workers will keep on monitoring nutrition status in a monthly frequency. Monitoring of risk factors occurs less frequently. The entire process makes the execution of the risk identification and intervention smoother and transparent while making data available for analysis and decision-making at all levels of hierarchy up to national level.

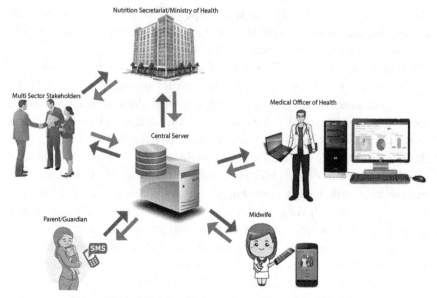

Figure 7.2 Information flow across multisector stakeholders.

2. Use of mobile technology in the public health sector: Asian Regional context

Mobile technology is mostly used in the clinical healthcare sector in multiple instances.

Yet, we observe several examples in the use of mobile technology for various applications in public health sector. In Indian context, a simple instance of sharing the mobile phone number of a community physician to be consulted over the phone round the clock revealed that people preferred this technology over traditional consultancies due to several factors [16]. This study suggested that 80% of the persons who used this service were males, which could be attributed to the fact that this opened up the avenue for a section of population who did not prefer conventional timings of consultations which overlapped with their working times. This was also supported by the fact that majority of the calls were made outside of office hours. Issues related to mental health and sexual disorders were of the most queried upon subject matters during these consultancies revealing patients' preference to discuss these matters, not in person with the healthcare provider in a traditional physical consultation setting. The article further revealed that 96% of the people wanted to continue with the service highlighting the high rate of acceptability even in the rural setting. There are examples in north Indian context in applying mobile technology for dissemination of health information for mother and new-born care mainly targeting women [17]. Thus, mobile technology could be utilized to reach sections of communities which are sometimes considered cornered in the local context.

It is, however, required to consider the ethical and personal concerns in using mobile technology for healthcare purposes since mobile devices are more linked to personal life of an individual. Yet, a study in rural area in Southern India revealed that nearly everyone participated in the study preferred receiving health-related communications in their mobile devices [18]. A randomized controlled trial conducted on pregnant ladies who received frequent SMS alerts on health education during pregnancy in Thailand suggested that mHealth interventions resulted in higher confidence levels and lower anxiety levels in pregnant ladies highlighting the efficacy of this technology to reach the extra mile following the antenatal care provided in clinics [19].

Use of mobile technology by field health workers in their routine activities are found at early implementational levels in few countries in the region. In Indonesia, Midwives Mobile-Phone project targets the use of mobile for better delivery of provision of maternal and infant care [20].

Another application of mobile technologies in public health sector which is increasingly seen in the region is to design information system which connects antenatal care of the mother and the new-born and immunization care of the child. This comes under the broader concept of Reproductive, Maternal, Newborn and Child Health (RMNCH) [21]. The combination of antenatal care and child healthcare information systems usually has components to record clinical information as well as to schedule clinic visits and to send reminders to enhance the coverage of services in the community [22].

Despite these widespread implementations, it has been revealed that minimal effort has been made in general in developing country perspective to provide theoretical understanding on adoption and effect of the technologies to produce measurable outcomes in healthcare or its delivery which significantly limits justifying the use of ICT and mHealth in formulating policy decisions [23].

3. Opportunities

3.1 Objectives

- Understand how mHealth contributes to enhancing the flow of information
- Understand the benefits of mobile technology to public health worker
- Identify how clients benefit from the use of mobile technology in public health

3.2 Content

District Nutrition Monitoring System (DNMS) is an award-winning innovative information system implemented by the Ministry of Health, Nutrition and Indigenous Medicine in Sri Lanka which uses smart mobile device to capture nutrition-related data at field level and share it across multisector stakeholders [24]. This novel concept of use of information and communication technology for multisector collaboration opens up whole range of opportunities. These will be discussed in following subcategories.

3.3 Information flow

The process creates a new dimension of opportunities to the traditional approach of healthcare delivery in a multisector setting. Data capture occurs at the field level, eliminating the delay of initiation of information flow. This speeds up the information exchange between stakeholders. Yet, the quality of data is not compromised. There are accuracy checks that are in place in mobile app itself. Fig. 7.3 depicts a screenshot of the mobile app where the field health worker captures nutrition parameters. He is prompted with the color code based on the WHO reference values for height and weight for the age of the child. These visual aids minimize data entry errors at point of data entry.

Once the data are captured at field level, these have to be supervised by the immediate supervising officers of public health midwives, the field level healthcare worker. Public health nursing sister is the supervising officer who checks the accuracy and validity of the records captured by the midwife. The corrected records will be synchronized with the midwife's device providing her a feedback on entered data. This establishes the accountability of data even at operational staff at grassroot level.

Sharing of data among multisector stakeholders is always challenging in a paper-based setup. The delay in transferring from paper to electronic system is a time-consuming exercise in a hybrid system. Thus, the process of

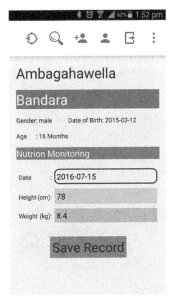

Figure 7.3 Data entry screen of mobile app with real-time color-coded validations.

sharing of data with multiple stakeholders is fast and better facilitated by use of mobile technology.

3.4 Healthcare worker is empowered

Sri Lanka boasts a well-established public health information system from field level to national level which has principally been a paper-based system. A salient drawback of this system is that it essentially makes the field health worker a data collector who feeds data to the broader national public health information system focused at strategic planning. District and operational level data managers such as public health doctors make use of analyzed data for planning at district level but rarely does this information transmitted back to field health worker who collects them. Use of mobile technology provides the field worker with the power of data analysis at fingertips which provides her with a feedback on collected data. The analytics provides her wisdom on issues to focus and assist her in planning her actions.

Availability of minimal data analysis and trend of nutrition information with graphical outputs in the mobile app enhances the accuracy of data capture and thereby improves the quality of data capture at field level. Visualizations such as growth charts in the mobile app provide a valuable opportunity for health education during field visits. Real-time analytics

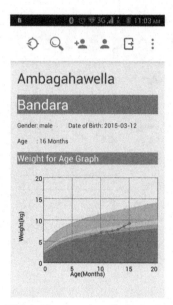

Figure 7.4 Weight for age growth chart on mobile app.

also enable the health worker to provide feedback on efforts of parents (Fig. 7.4).

Traditional view of field health worker carrying a pile of files and paper records in field visits has been a negative factor for job satisfaction and attraction for new recruitments for field health vacancies Fig. 7.5).

The new role of "smart field health worker" equipped with a smart mobile device has created a value addition to the job performed (Fig. 7.6). Quality of field-level data capture could be enhanced by validations and visual aids available in the data capture app.

3.5 Benefits to clients

The traditional model of data capture was on a paper-based format where the collected data was mainly used by the field health worker for monitoring the nutritional status. Sharing of data with superior officers as well as with multisector stakeholders was slow and time-consuming. Thus, care at individual child level was based much on partial judgment of the healthcare worker with minimal inputs from multisector stakeholders. Use of mobile technology enables fast capture and sharing of data which ultimately results in better care at client level. Use of mobile app enhances the quality of data capture which in turn benefits the type of care provided to the child by the

Figure 7.5 Public health midwife in a rural area in Sri Lanka in field work.

Figure 7.6 Public health midwife entering data at a field health clinic.

public health midwife. Due to the fact that analyzed data are available with the field health worker, she can provide health education and care based on informed decisions.

Feedback obtained from parents of children presented to field weighing centers revealed interesting findings. The parents felt delighted that the field health workers use the latest technology in public health—related work. The general perception of government services in minds of public tend to be quite negative. Since use of mobile technology is a trending practice in private sector, the new trend induced by the DNMS mobile application was highlighted as a positive. In addition, they were quite keen to get the data accessed using their mobile devices as a personal health record, which was not implemented up till now.

4. Challenges

4.1 Objectives

- Outline the hardware-related challenges in mHealth Implementation
- Understand how software-related factors become a challenge
- Identify the connectivity-related issues in mobile implementations
- Understand the importance of capacity building

4.2 Content

The process of implementing the mHealth solution has not been smooth nonetheless and was associated with a number of challenges. These will be discussed under following headings.

4.3 Hardware

Initial capital cost on procurement of mobile devices was a major constraint during the pilot and scaling up. The issue got further complicated since the project was initially funded by the development organization for the pilot phase. There was no proper funding allocation from a ministry of health budget. Hence, agreement with mobile phone vendors had to be carefully drafted to incorporate extensive service agreement lasting for a few years until the ministry of health budgetary allocation was approved. Another area that must be planned is the requirements for the future. It is mandatory to speculate the device requirement for next few years which also should count the need for device replacement due to loss or malfunctioning as well as recruitment of new cadre.

Selecting the type and model of mobile device with financial constraints was always a concern. There will always be pressure from districts as well as planning staff from ministry and implementing partners to utilize the available budget to purchase maximum number of devices to expand the implementation. But it must be emphasized that the decision of type of device and the model has to be decided based on technical grounds. Based on our experience and feedback from users, the factors which contribute to this decision are as follows:

- The length and complexity of data entry forms—if lengthy forms with a large number of checks, validation rules tablet mobile devices are well-suited
- Portability of device—if portability is a major concern, the smaller the screen size is better
- Whether the device replaces the field health worker's mobile device—if this is expected to happen, provision of a smart mobile device of small screen size is preferred
- If battery life is a concern—usually tablet mobile devices with larger screen sizes have batteries which can last longer
- Cost per device—usually mobile phones are quite cheaper compared to tablet mobile devices in Sri Lanka

There was a trade-off between screen size and the portability as well as hardware resources of the mobile device and the cost of the device. The channel of aftercare service had to be designed from the national level up to field level for maintenance and service of mobile devices. The strategy was to provide repair services through a company which had country-wide branches which could provide services to end-users. Replacement of devices due to physical damages and at end of lifetime was always interesting and needed meticulous planning.

Setting up of information system required a server with a significant level of resources. This was made possible with the procurement of a server to the ministry of health with the funding of a development partner which saved significant recurring costs of web hosting due to the lack of availability of quality web hosting facilities within the country, which is usually the case in developing countries such as Sri Lanka. This becomes worse by the fact that legislation and policies of the ministries mandate the data to be hosted locally within the country. It is, however, required to assign funding to replace and upgrade server facilities periodically from the ministry budget which usually is an area not concerned by implementers in initial stages.

4.4 Software

Software development for mobile applications is always challenging compared to web app development [25]. Limited screen size and the resolution forces the designer to only include the content which is essential in the interfaces. The thin balance between detail and the clarity of the interface is what tests the designing ability of the development team. A number of interfaces have to be limited to enhance user experience. The orientation of the device whether portrait or landscape has to be considered per each interface due to limited screen area. In mobile applications, icons and colors replace most of the textual content that are usually available in web apps. Hence, proper use of icons and colors could save significant amount of screen area. Another factor that has to be considered when engaging with graphics is the availability of resources in the mobile device. Graphics used in mobile applications have to be compatible with the relatively modest memory and processing speed available in devices distributed to field health workers.

The mobile application needed to support offline data entry and analysis to be optimally used in areas of low network coverage. This has been a top priority functionality the mobile app was based on. The field health workers are able to collect and store data in the app offline, which could be synchronized to the online server when mobile Internet is available. The solution was implemented in a district of Sri Lanka which had difficult terrain and poor mobile network reception due to mountains ranged across most parts of the district. Due to offline support of the mobile application, we had no complaints from the end-users on issues in not being able to use the app in the field.

Another issue that is still a challenge is the version upgrade and user management. Frequent feedback is obtained from field-level staff on interfaces and functionality of the mobile app. This is one major aspect driving the development of the mobile application. In addition, mobile app needs updating based on release cycles of DHIS2 (District Health Information Software 2) software platform where the data are synchronized to Ref. [26]. Thus, new versions of the mobile app are developed and distributed to the end-users by the use of Google Play Store, which is the current mode of distribution of the mobile app. A practical issue encountered with this strategy is the loss of control over the distribution. End-users install the application when they desire and sometimes it is not possible to determine the user base who are on the latest version of the app. Therefore, a mobile device management solution [27] becomes handy to apply this situation. This could bring about better control over the version of app utilized by the field health staff and can also be utilized to control the use of mobile device. This is an aspect currently considered by the implementers. A major challenge is the cost associated with a subscription of a mobile device management software which usually incurs a significant recurrent budget. Since the districts in which initial implementation of the system occurred, had mobile devices purchased from the ministry of health, implementing a mobile device management solution will not be deemed any major concerns from end-users. Yet, if such solution is to be implemented in devices procured by another organization or on personal devices of the end-users, it will be likely to be met with resistance from end-users.

The mobile application needs to be updated to be on par with the release cycles of DHIS2 software, the platform on which the system is based on. In addition to this, with more data collected on the system, the mobile app needs to be updated to enhance synchronization time as well as user interfaces with updates based on feedback provided by end-users.

4.5 Network management

Costs for mobile data were always a concern, and regulating mobile data at the level of end-users was another challenge. During initial period of implementation, midwives were provided with a monthly incentive to cover costs of mobile data. But this was never proven to be sustainable with diminishing donor funding. In spite of having sessions on managing the data quota during the training program, this was not proven to be successful. There were also suggestions to come into a service agreement with a single mobile network operator on subsidized rate for mobile data. This was proven to be ineffective since entire country is not equally covered by all mobile operators and hence there was always the chance that selected mobile network operator did not have satisfactory coverage to be utilized by all health workers. A better approach which is currently being discussed at national level is to come into terms with mobile network providers to subsidize the mobile data charge for requests getting connected to IP addresses of servers where nutrition data servers are hosted.

The mobile data management plan has to be executed along with a mobile device management plan in which national core team could regulate access of mobile apps to the Internet. This, however, may be deemed acceptable in Sri Lankan setting where the ownership of the mobile devices lies with the ministry of health, but if the mobile app is installed on personal mobile devices of end-users, this may not be feasible to implement.

4.6 Capacity building

End-user training and capacity building required a considerable amount of time and resources. This necessitated capacity-building framework to be implemented at intermediate levels of hierarchy in information flow and not just the end-user level. The capacity building required training on public health information capture and analysis as well as technology-related areas on troubleshooting and use of ICT.

Initial training programmes for end-users were conducted at Medical Officer of Health (MOH) level which is the smallest cluster of field-level health workers, managed by a public health doctor. One full day in-service training program was conducted to public health midwives at respective MOH offices to orient them on proper use of mobile technology and the mobile app (Fig. 7.7).

The mobile application was implemented in the field following the one-day training program, yet support was required for routine troubleshoot. The team of implementers were available over the phone for this task. It was, however, realized soon enough that this method of remote troubleshooting was proven to be ineffective for field-level issues. A series of capacity-building activities were then planned which mainly targeted ICT staff at district level. ICT officers at district health offices were trained on mobile device—related troubleshooting as well as the mobile application. In addition, another training program was arranged for selected few midwives from each MOH area who were above average in technological competency. The training program was focused more on the troubleshooting perspective. This method of peer troubleshooting was particularly effective in providing end-user assistance mainly due to the facts that there was less barrier in approaching help offered from a coworker and the fast response time since they were readily available in physical vicinity.

Figure 7.7 Training of end-users at field offices.

4.7 Monitoring and postimplementation support

A key to successful implementation lies mostly with the post-implementation effort exerted by the implementers. It is the usual practice to plan and target implementation plan up to the implementation at field level. Yet, the bulk of work is required to sustain the efforts once the system is in use. Refresher training, updating data collection formats based on feedback, and provision of continuous finances are some aspects which have been recognized in developing country perspective [28]. These are applicable to the Sri Lankan context as well. In addition it was required to operationalize the system by integrating the mobile application and the monitoring of field health worker performance in monthly conferences held at MOH's office. The meetings are headed by the public health doctor who is in charge of the area. It was expected for them to display the dashboards of the system with key nutrition and health worker performance indicators to identify progress of malnutrition in the area of concern and to assess the work of the field health workers. This can be used for rewarding the end-users with good performance and to identify gaps and encourage health workers whose performance are below average.

The monitoring and review should happen at all levels of health hierarchy for the successful operationalization of the system. This necessitates there should be review activities happening at national level which should cascade to provincial and district levels which in turn will encourage public health doctors to perform monitoring of data collection of public health midwives.

Implementation of mHealth projects at field level in developing countries requires proper planning and stepwise implementation. It is not uncommon to find implementations scaling-up in an overambitious path. Usually these systems tend to collapse overtime since each step in implementing the mHealth system requires setting up infrastructure and local capacity building without which the system fails to sustain. Thus, the exercise involves constant challenges with rewarding outcomes.

In summary, this chapter looks at the use and applicability of mHealth for public health monitoring and intervention in low- and middle-income countries. It discusses the context of mHealth in the Asian region and specifically focuses on the use case in domain of nutrition in Sri Lanka and elaborates the opportunities and challenges resolved around this project.

mHealth open the immense opportunities for the LMICs. mHealth tools speed up the information exchange from the field level all the way up to the national level while ensuring the quality of data. This is achieved by the possibility of applying validation checks and other quality control mechanisms at the point of capture of data. It even provides real-time analytics to the field health worker. Real-time sharing of data across multiple stakeholders enhances information exchange as well as evidence-based decision-making. Feel health worker is empowered by use of mHealth tools. There is feedback received for the data captured. Use of technology and smart devices adds value and brings social recognition to a job role which was considered otherwise merely as a field worker. Availability of minimal data analytics and related information facilitate decision-making for the field health worker. Use of smart technology by the healthcare workers which empowers decision-making and account-ability overall enhances the quality and the responsibility of care provided to the public which is seen as a significant benefit provided to the public equivalent to the type of care available in developed countries. In addition, use of technology brings more value to the state health sector which was exclusively enjoyed by the private sector in past.

Nevertheless, implementation of mHealth projects involves many challenges. Hardware-related challenges involve many dimensions. The significant cost associated with servers as well as data capture devices in the implementation process in most of the cases. This sometimes leads to purchase of low quality devices which usually involves durability concerns. Information system hosting and maintenance of servers also become challenging due to the low resources available. Designing of mobile applications field-level use requires proper planning and feedback from the end-users. It might need several modifications before the final product. Capturing of minimal data, use of icons and images, and offline function-ality are major concerns in designing such tools. Ensuring sustainability and maintainability of the information system has to be thought of during initial planning. Distribution of the mobile tool and the versioning are also other factors that need to be thought of. Mobile data collection is always associated with mobile Internet. Costs and maintainability of mobile data packages is to be ensured for continuity of the system as well as for compliance from end-users. Capacity building is the single most important thing for the implementation of ICT-related technologies. Proper measures have to be taken to ensure training and capacity building at all levels involved from national level up to the field level. Policy decisions have to be taken to ascertain continuity of capacity building. Monitoring and postimplementation support are unnecessary to enhance the quality of implementation as well as sustainability.

References

[1] World Health Organization. mHealth: new horizons for health through mobile technologies. 2011. Retrieved from: http://www.who.int/about/.

[2] World Health Organization. WHO | telehealth. 2016. Retrieved June 30, 2019, from: https://www.who.int/sustainable-development/health-sector/strategies/telehealth/en/.

[3] Kim D-K, Yoo SK, Park I-C, Choa M, Bae KY, Kim Y-D, Heo J-H. A mobile telemedicine system for remote consultation in cases of acute stroke. J Telemed Telecare 2009;15(2):102−7. https://doi.org/10.1258/jtt.2008.080713.

[4] Rechel B, McKee M. Facets of public health in Europe. European observatory on health systems and policies series. 2014. Retrieved from: https://scholar.google.com/scholar_lookup?title=Facets+of+Public+Health+in+Europe.+European+Observatory+on+Health+Systems+and+Policies+Series&author=B.+Rechel&author=M.+McKee&publication_year=2014&.

[5] Grady A, Yoong S, Sutherland R, Lee H, Nathan N, Wolfenden L. Improving the public health impact of eHealth and mHealth interventions. Aust & N Z J Public Health 2018;42(2):118−9. https://doi.org/10.1111/1753-6405.12771.

[6] WDI - classifying countries by income; n.d. Retrieved November 2, 2019, from: https://datatopics.worldbank.org/world-development-indicators/stories/the-classification-of-countries-by-income.html.

[7] WHO | Definition of regional groupings. WHO; 2017. Retrieved from: https://www.who.int/healthinfo/global_burden_disease/definition_regions/en/.

[8] Puchalski Ritchie LM, Khan S, Moore JE, Timmings C, van Lettow M, Vogel JP, et al. Low- and middle-income countries face many common barriers to implementation of maternal health evidence products. J Clin Epidemiol 2016;76:229−37. https://doi.org/10.1016/J.JCLINEPI.2016.02.017.

[9] Lopéz DM, Blobel B. mHealth in low- and middle-income countries: status, requirements and strategies. Stud Health Technol Inf 2015;211:79−87. Retrieved from: http://www.ncbi.nlm.nih.gov/pubmed/25980851.

[10] GSMA mHealth tracker | mobile for development; n.d. Retrieved November 2, 2019, from: https://www.gsma.com/mobilefordevelopment/m4d-tracker/mhealth-deployment-tracker/.

[11] Wallis L, Blessing P, Dalwai M, Shin SD. Integrating mHealth at point of care in low- and middle-income settings: the system perspective. Glob Health Action 2017;10(Suppl. 3):1327686. https://doi.org/10.1080/16549716.2017.1327686.

[12] GMS Association. Digital identity country report:Sri Lanka. 2019. Retrieved from: https://www.gsma.com/mobilefordevelopment/wp-content/uploads/2019/02/Digital-Identity-Country-Report-Sri-Lanka.pdf.

[13] Understanding the true cost of malnutrition; n.d. Retrieved November 2, 2019, from: http://www.fao.org/zhc/detail-events/en/c/238389/.

[14] Galgamuwa LS, Iddawela D, Dharmaratne SD, Galgamuwa GLS. Nutritional status and correlated socio-economic factors among preschool and school children in plantation communities, Sri Lanka. BMC Publ Health 2017;17(1):377. https://doi.org/10.1186/s12889-017-4311-y.

[15] Alderman H, Elder L, Goyal A, Herforth A, Hoberg Y. Improving nutrition through multisectoral approaches. 2013. Retrieved from: https://www.popline.org/node/566913.

[16] Bali S, Singh AJ. Mobile phone consultation for community health care in rural north India. J Telemed Telecare 2007;13(8):421−4. https://doi.org/10.1258/135763307783064421.

[17] Kumar N, Anderson RJ. Mobile phones for maternal health in rural India. In: Proceedings of the 33rd annual ACM conference on human factors in computing systems - CHI'15. New York, New York, USA: ACM Press; 2015. p. 427−36. https://doi.org/10.1145/2702123.2702258.

[18] DeSouza SI, Rashmi MR, Vasanthi AP, Joseph SM, Rodrigues R. Mobile phones: the next step towards healthcare delivery in rural India? PloS One 2014;9(8):e104895. https://doi.org/10.1371/journal.pone.0104895.

[19] Jareethum R, Titapant V, et al. Satisfaction of healthy pregnant women receiving short message service via mobile phone for prenatal support: a randomized controlled trial. Si.Mahidol.Ac.Th. J Med Assoc Thai 2008:458−62. Retrieved from: https://www.si.mahidol.ac.th/th/publication/2008/vol91_no.4_458_1404.pdf.

[20] Chib A, Lwin MO, Ang J, Lin H, Santoso F. Midwives and mobiles: using ICTs to improve healthcare in Aceh Besar, Indonesia1. Asian J Commun 2008;18(4):348−64. https://doi.org/10.1080/01292980802344182.

[21] Black RE, Walker N, Laxminarayan R, Temmerman M. Reproductive, maternal, newborn, and child health: key messages of this volume. Reproductive, Maternal, Newborn, and Child Health: Disease Control Priorities. 3rd ed., 2. The International Bank for Reconstruction and Development/The World Bank; 2016.

[22] Kaewkungwal J, Singhasivanon P, Khamsiriwatchara A, Sawang S, Meankaew P, Wechsart A. Application of smart phone in "Better Border Healthcare Program" a module for mother and child care. BMC Med Inf Decis Making 2010;10(1):69. https://doi.org/10.1186/1472-6947-10-69.

[23] Chib A, van Velthoven MH, Car J. mHealth adoption in low-resource environments: a review of the use of mobile healthcare in developing countries. J Health Commun 2015;20(1):4−34. https://doi.org/10.1080/10810730.2013.864735.

[24] District Nutrition Monitoring System (DNMS) | WSA; n.d. Retrieved November 2, 2019, from: https://www.worldsummitawards.org/winner/district-nutrition-monitoring-system-dnms/.

[25] User interface design for the mobile web − IBM Developer; n.d. Retrieved July 14, 2019, from: https://developer.ibm.com/articles/wa-interface/.

[26] DHIS2; n.d. Retrieved April 22, 2018, from: https://www.dhis2.org/.

[27] What is mobile device management? Webopedia definition; n.d. Retrieved July 16, 2019, from: https://www.webopedia.com/TERM/M/mobile_device_management.html.

[28] Kasambara A, Kumwenda S, Kalulu K, Lungu K, Beattie T, Masangwi S, et al. Assessment of implementation of the health management information system at the district level in southern Malawi. Malawi Med J 2017;29(3):240−6. https://doi.org/10.4314/mmj.v29i3.3.

CHAPTER 8

How to use the Integrated-Change Model to design digital health programs

Kei Long Cheung, PhD[1], Santiago Hors-Fraile, MSc[2],
Hein de Vries, PhD[3]

[1]Department of Health Sciences, Brunel University London, London, United Kingdom; [2]Salumedia Labs, Seville, Spain; [3]Professor in Health Communication at the Department of Health Promotion, CAPHRI Public Health, Maastricht University, Maastricht, The Netherlands

Contents

The overall goal of this chapter is that the reader gains knowledge, understanding, and skills regarding a pragmatic methodology to design tailored digital health (dHealth) programs, using the Integrated-Change Model (I-Change Model). Several learning outcomes can be derived from this overall goal, with varying complex levels of understanding: remembering, understanding, applying, analyzing, and evaluating.

After reading and conducting exercises of this chapter, the reader should be better able to

1. Describe several sociocognitive theories to understand health behavior
2. Define important determinants of behavior

Digital Health
ISBN 978-0-12-820077-3
https://doi.org/10.1016/B978-0-12-820077-3.00008-0

143

3. Describe how sociocognitive determinants can affect health behavior and health outcomes
4. Explain the integration of theories to understand and predict behavior, applying the I-Change Model
5. Outline the pathway of change of a behavior change intervention
6. Describe how qualitative and quantitative research methods feeds into the intervention development

Unhealthy lifestyles, such as tobacco smoking and excessive alcohol consumption, lead to lifestyle-related diseases. These in turn negatively impact health and quality of life at the individual as well as the societal level. In order to change unhealthy lifestyles into healthier ones, it is vital to understand why the individuals engage in a certain behavior. In this chapter, we will discuss several theoretical approaches to understand health behavior, and in particular the I-Change Model. A pragmatic approach to build health interventions will be discussed, with an illustration of applying the I-Change Model to build a computer-tailored smoking cessation intervention in the subsequent chapter. Furthermore, a new approach to tailoring, using a data-driven algorithm, will be discussed.

Exercise 1: Digital flyer about tobacco smoking in the Farmerlands

In a small country far away, named the Farmerlands, more funding can be provided to improve the health of its citizens. This is needed as national health monitoring reports have shown that there is a steep increase in heart diseases, cancers, and COPD. Public health researchers conclude that tobacco smoking is a major problem among the adult population. The idea now is to diffuse a digital flyer, meaning a static website and an e-mail, with the aim to enhance knowledge on tobacco smoking among citizens of the Farmerlands. Do you think this will be effective to change smokers to nonsmokers?

---

1. Theory to understand health behavior

When designing health interventions, it is important to understand health behavior, such as tobacco smoking, physical activity, or alcohol consumption. The identification of potential factors that influence health behavior (i.e., behavioral determinants) has been derived from many theoretical models from different disciplines, such as psychology, sociology, and anthropology. One type of theoretical models entails sociocognitive models, describing the interplay of social and cultural factors on the development of behavior and ideas about behavior, also referred to as beliefs or cognitions. Sociocognitive models help one understand the associations among the determinants of behavior change and how these determinants influence behavior. We will now briefly describe a few relevant sociocognitive models in order to understand health behavior.

A highly influential sociocognitive model is Bandura's *Social Cognitive Theory* [1]. Bandura outlines not only that a person's behavior is influenced by personal factors (e.g., belief in one's capabilities to perform a specific behavior) and environmental factors (e.g., social influences that can influence the individual's ability to perform a specific behavior) but also that behavior influences the personal factors and environmental influences. In other words, a person's behavior is continuous and in reciprocal interaction with other personal and environmental influences. The environmental factors (or also referred to as situation) pertain to the factors that are external to a person. A person's environment provides the social and physical context in which a person functions, which can reinforce or punish particular behaviors. Hence, social factors are important to consider when understanding behavior change, such as role models. Personal factors refer to the abilities of a person to learn from behavior through own experiences or observation of other's behavior. The latter is referred to as vicarious learning, which is a type of learning that is learned indirectly through observation (e.g., whistle after watching someone else shape their lips and expel air). The behavioral factor refers to the capacities a person has to be able to actually perform a behavior, for instance, specific skills or intellectual abilities.

Bandura therefore outlines that learning occurs in a social context with a dynamic and reciprocal interaction of the person, environment, and behavior. This interaction is known as reciprocal determinism [2]. Reciprocal refers to the mutual influence between these three factors, suggesting that the learner is not a passive recipient of information. Three factors from the Social Cognitive Theory (see Fig. 8.1) that drive an

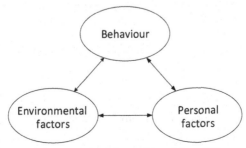

Figure 8.1 Social cognitive theory [1].

individual's behavior which have become very popular among behavior scientists are (1) outcome expectations, (2) perceived modeling, and (3) self-efficacy expectations.

1. Outcome expectations refer to a person's expectations about the outcomes or consequences of their behavior.
2. Modeling refers to the perception of others performing a particular behavior (and thus an environmental factor), which leads to the imitation of it when the behavior results in positive outcomes. Alternatively, modeling can also result in avoidance of the behavior when negative outcomes are perceived.
3. Self-efficacy refers to a person's belief (thus a personal factor) whether he or she is capable of performing the behavior.

The *Health Belief Model* (HBM) [3] is an influential model to understand health behavior, stemming from the 1950s with many applications in public health. HBM attempts to explain and predict health behaviors by focusing on the attitudes and risk beliefs of individuals. The HBM stipulates that behavior change is the result of the perceived threat and net benefits (see Fig. 8.2), which are the results of four factors: perceived susceptibility (i.e., an individual's belief of the chances of getting a condition), perceived severity (i.e., an individual's belief of how serious a condition and its consequences are for him or her), perceived benefits (i.e., an individual's belief in the

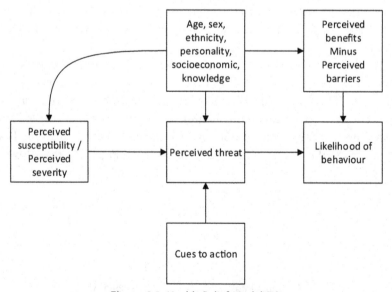

Figure 8.2 Health Belief Model [3].

efficacy of the health behavior to reduce risk or seriousness of impact), and perceived barriers (i.e., an individual's belief regards the disadvantages of the health behavior) [3]. Cues to action, which could be internal and external cues (such as pain, internal), would facilitate the likelihood of behavior change by tapping into the perceived threat. In later models, self-efficacy was added to an adapted and extended HMB.

Another popular social cognitive model concerns the *Theory of Reasoned Action* [4,5] that stipulates that behavior is predicted by intentions to perform the behavior. The intention (i.e., an individual's readiness to perform a specific behavior) is in turn is predicted by the attitudes of a person toward that behavior (i.e., the perceived outcomes of the behavior and the value placed upon the outcomes) and the subjective or social norms concerning that behavior (i.e., the perception of how others would view the behavior). Thus, attitudes entail the perceived beliefs about the outcomes of the behavior (e.g., physical activity will help me to remain slim). Social norms refer to the perceived beliefs of others about the behavior (e.g., my partner believes that I should be more physically active). Later Bandura's concept of self-efficacy from Social Cognitive Theory [6] was added although referred to as perceived behavioral control in the *Theory of Planned Behaviour* [7] (see Fig. 8.3).

Other sociocognitive models are important as well, such as the *transtheoretical model* [8], stipulating that behavior change is a gradual and dynamic process in which individuals move through several stages of change. Another example is *goal setting theory* [9], outlining that behavior

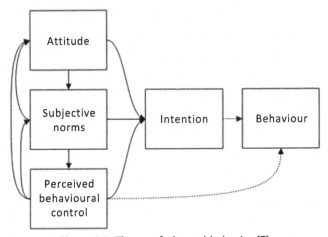

Figure 8.3 Theory of planned behavior [7].

can be explained by the development of plans, designed to motivate and guide an individual to reach their goal, such as performing a specific behavior. In conclusion, there are several sociocognitive theories emphasizing different underlying mechanisms to behavior change; each model with their own strengths and weaknesses. Hence, in order to better understand the mechanism to behavior change, it is relevant to critically test and integrate constructs and theories (which has already been reflected in the theories discussed). Many ideas of the discussed theories are incorporated in a comprehensive integrated model, which we will now discuss.

2. Integrated-Change Model

The *I-Change Model*, or the Integrated Model for explaining motivational and behavioral change [10], integrates ideas of Bandura's Social Cognitive Theory [1], the HBM [11], Theory of Planned Behaviour [7], Prochaska's Transtheoretical Model [8], and Goal setting theory [9]. The I-Change Model (see Fig. 8.4) stipulates three phases of behavior change: (1) premotivation or awareness, (2) motivation, and (3) postmotivation or action.

In order to adopt a healthy behavior, such as nonsmoking via smoking cessation, the first phase is that one becomes aware of the risks of a specific behavior (e.g., tobacco smoking) and recognizes that one is actually engaging in this behavior. *Awareness* consists of the factors such as

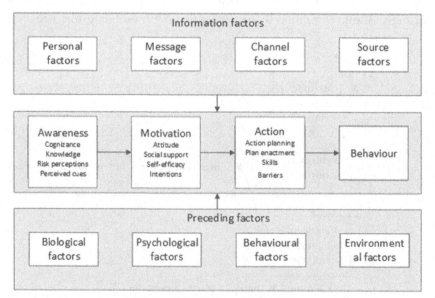

Figure 8.4 I-Change Model [10].

knowledge, risk perception, cognizance, and perceived cues. Knowledge represents the understanding of the facts of the risk behavior and the how to perform the health behavior. Risk perception is the perceived susceptibility and the perceived severity of a health threat. Cognizance is adequately recognizing that one is performing a risky health behavior and to which extent. For instance, many smokers underestimate the number of cigarettes that they are smoking, a phenomenon also occurring among users of both tobacco and e-cigarettes Awareness about a specific problem can also occur via the perception of cues. These can be internal (e.g., coughing in smokers) or the neighbor mentioning his lung cancer diagnosis or attention to specific antismoking campaigns. If someone is not aware of being at risk, this person will interpret health communications differently from someone who is aware of being at risk. For instance, smokers that belief that tobacco smoking is not a risky behavior, because their grandfather smoked until the age of 80, are more likely to discard health communication messages about smoking, than those who are aware of smoking being a risk behavior.

When an individual is aware of a health risk behavior and the personal risks, the next phase is to enhance the *motivation to change*. The motivation to change is reflected by the intention to change (i.e., overall motivational state to change a specific behavior). The determinants of intention are attitude, social influences, and self-efficacy, which we also have discussed above. Attitude refers to a person's overall evaluation of the consequences of a health behavior and thus concerns the advantages and disadvantages of a health behavior. Social influences refer to the perceived influences from others regards a health behavior, reflected by the social norms, modeling, and support or pressure. Self-efficacy is a person's perception of the own capacity to perform a particular behavior in different situations. After motivating a person to adopt a new behavior, this intention does not always translate to actual behavior change. A high motivation, or intention, to adopt a (health) behavior does not always translate to actual behavior change.

When a person is motivated to change, this person can move to the *action phase* where the focus is on how to translate intentions into actions. This translation is dependent (again) not only on a person's level of self-efficacy but also on whether action plans are made and enacted. Action planning entails action planning, preparatory planning, coping planning, and plan enactment. For instance, when a person has the intention to quit smoking, a person needs to make plans on how to do this (e.g., using counseling, pharmacotherapy, eHealth). Then specific action plans are made (e.g., I will quit smoking using counseling; I will quit smoking using pharmacotherapy). Action planning are specific planned behaviors that ultimately lead to the overall health behavior. Preparatory planning entails specific plans to help undertaking the behavior change attempt. For instance, when a person plans to quit smoking next week, a preparatory plan could be to take away all ashtrays in the house. When engaging in specific change attempts, a person is likely to encounter various difficulties; difficulties that are related to the specific situations (e.g., how to remain quit when: gaining weight; when feeling stressed, etc.). It is important that a person will develop specific coping plans for these difficult situations (e.g., when I feel stressed, I will drink a glass of water, or go for a walk, or listen to music). Hence, coping planning are strategies to handle difficult situations that may endanger maintenance of the newly required health behavior. People can easily make plans, yet the realization of these plans is vital, a phenomenon that is referred to as plan enactment in the I-Change Model. Lastly, a person can plan various actions but does need specific skills to realize these plans and to be able to cope with barriers that may hinder the behavior.

Exercise 2: Limited effects of the digital flyer

The answer to exercise 1 is: probably not! It turns out that the digital flyer in the Farmerlands had extremely limited effects on eliciting behavior change and sustained smoking cessation. Explain why!

---

3. Digital health and tailoring

Now we have a better understanding of the underlying mechanism to health behavior change, so we will discuss the opportunities of digital health (dHealth) and how we can apply this knowledge to create tailored dHealth interventions. dHealth is an umbrella term for the usage of digital technology to support health [12] and thus can refer to a great variety of interventions, from static websites to interactive apps. dHealth is one approach to address risky lifestyle behaviors, with the potential to be highly cost-effective [13,14] as these software-based solutions can be scaled with minimum or none costs. dHealth interventions have shown effectiveness to change various behaviors, such as physical activity, alcohol consumption, condom use, dietary intake, and tobacco smoking [13], and can reach a broad audience due to the incredible rise of the Internet on a global scale. So how do we develop effective dHealth interventions? Not all variations of dHealth are equally effective. One approach of dHealth that seem especially effective is the use of computer-tailoring [13,15]. Computer-tailoring means adjusting intervention (health) messages or components to the specific user's characteristics via digital technology. Tailored dHealth uses computer-tailoring technology to deliver motivational health information personalized on user demographics (e.g., age and gender) and sociocognitive profile (e.g., attitudes and risk perceptions). When looking at systematic reviews of dHealth interventions to enhance smoking cessation for instance, interventions that are theoretically based tailored to the individual have the highest effect sizes [13,15]. These tailored interventions rely on theories to effectively tailor to the individual. The I-Change Model explains the underlying mechanism to how behavior change can be elicited and have performed well in tailored dHealth programs [14,16—18]. For instance, a Dutch systematic review showed that all effective dHealth interventions are tailored using the I-Change Model as framework for its logic of change [13].

4. Pragmatic methodology to design digital health

We have briefly discussed the theoretical backbone to understand behavior change and that theoretically derived interventions that tailor the provision of health messages to individual needs are most effective. In order to develop computer-tailored interventions, several steps are relevant to address. Various publications can be found on this process (see, e.g., De Vries and Brug [19]; Dijkstra and De Vries [20]). In line with guidelines on how to design evidence-based health promoting programs [21—23], the journey starts with a needs assessment. This is the process before the health promoter thinks about solutions. This is an exploration of the problem: what behavior is maintaining the health problems? What is the target population, and what other people are involved? General frameworks for intervention development (e.g., Behaviour Change Wheel and Intervention Mapping) are also relevant to guide the overall program development [21]. These planning protocols, such as Intervention Mapping, provide detailed guidance for the needs assessment. When time and resources are limited, and when it is clear which health behavior needs to be changed, a more pragmatic approach can be taken. The pragmatic methodology to design tailored dHealth entails broadly four steps: (1) choice of the theoretical model; (2) goal and objectives of the tailored dHealth program; (3) identification of salient beliefs; and (4) design the program content and algorithm.

4.1 Choice of the theoretical model

The selection of a theoretical model drives choices as to which factors will be addressed. For instance, the HBM does not discuss social norms, social modeling, and social support. The Theory of Planned Behavior does not address action planning. Hence, an informed choice is needed concerning which theoretical model is most appropriate for understanding and changing a particular health behavior. In this chapter, we use the I-Change Model, focusing on awareness, motivation, and action.

4.2 Goal and objectives of the computer-tailored intervention

Next, choices need to be made concerning the focus of the computer-tailored intervention pertaining to the overall goal and specific behavioral determinants that we want to address. Do we want to target persons already motivated to change? In this case, we can then focus on determinants such as self-efficacy, action planning, and skills development. Yet, different objectives are needed when the focus is to make people aware or to motivate people or a combination.

4.3 Identification of salient beliefs

As beliefs influence behaviors, it is needful to explore the beliefs of the target population regarding the health behavior. Some beliefs are more related to certain health behaviors than others. Thus, the relevant beliefs (and their related theoretical constructs) need to be captured, and their link with the actual health behavior.

4.3.1 Literature research

As resources and time are limited, the first step is to dive in the literature, exploring what is already known regarding the perceptions of your target population regards the behavior of interest, such as smoking cessation. Previous research in a similar context provides a good starting point to aid the information you need. Some studies do provide the information that may be useful for intervention development, the identification of the salient beliefs. For instance, Nierkens et al. provided insights into the salient beliefs of smoking cessation in Turkish and Moroccan immigrants in the Netherlands using the I-Change Model [24], or very specifically, Cheung et al. showed which beliefs are important to explain the intention to use an economic decision support tool for stakeholders of tobacco control [25]. However, in many cases, the literature does not provide the guidance, or the setting of the evidence may not apply to the context of the to-be-designed intervention. We then turn to formative research, using a two-step approach, starting with qualitative research and followed by quantitative research.

4.3.2 Qualitative research: exploring relevant beliefs

We need to explore the potential important beliefs of the target population, surrounding relevant determinants of behavior. Explorative open-ended questions can be based on the I-Change Model, addressing knowledge, risk perceptions, cognizance, attitude, perceived social norms, perceived social support, perceived social modeling, self-efficacy, and action planning. For this purpose, qualitative research methods can be used, for instance, via focus groups or interviews. Interviews are often employed to provide insights of the beliefs that may be paramount. For example, alcohol consumption beliefs could be understood through interviews of people who abuse from it, and from those who do not. Once these beliefs are extracted, we will be able to choose the matching behavior determinants to be acted on in our behavioral change model so that they foster the healthy behavior.

4.3.3 Quantitative research: identifying the salient beliefs

After having collected the set of potentially relevant beliefs and determinants influencing the health behavior, we have to identify the beliefs that will be more influential for our program of health promotion. These beliefs will explain the differences in the healthy and nonhealthy behavior in our population. Quantitative methods can be employed to identify these salient beliefs. Surveys studies are often applied to compare healthy and nonhealthy behavior via statistical analysis. We can translate the beliefs identified in our previous qualitative step into statements to be assessed in the surveys. An example of such a statement would be "I believe that smoking will help me to lose weight." This example would be an attitudinal belief, for which we will ask smokers and nonsmokers to rate it on a scale, such as a 7-point Likert scale. Statistical analyses, such as ANCOVA analyses, can then be used to identify the significant differences between smokers and nonsmokers. The significant beliefs are the salient beliefs that can explain the difference between persons performing the healthy behavior and persons who perform the nonhealthy behavior.

4.4 Design the program content and algorithm

Now the intervention design can be created, which is a blueprint outlining the logic of change (the theoretical pathway, determinants, and the specific beliefs to be addressed), and the intervention design also involves planning how intervention components can impact these beliefs in order to enhance the desired behavior change.

4.4.1 General idea and intervention components

In this process, it is important to involve all relevant stakeholders in a planning group, including the target population and the implementers, to enhance the match between stakeholder needs and the intervention characteristics. This will optimize the adoption and implementation of the intervention. The first step in the design is to discuss within the planning group what the general idea and components of the intervention are and how and in what sequence the intervention components are being delivered.

4.4.2 Create change objectives and match with change methods

The second and highly important step is to consider each salient belief to be an objective of change and to match change methods to reach the specific objective. In other words, use the salient beliefs to formulate program objectives, and consider how you are reaching those objectives. For instance, when quantitative analyses show that the belief "I think that smoking helps me to relax" is a salient belief, thus a belief that can statistically explain the difference between smokers and nonsmokers, the designer (i.e., the health promoter designing the intervention) can consider this belief to be a change objective of the dHealth intervention. That is to convince the smoker that smoking actually increases anxiety and tension. What health messages does the dHealth program have to provide to smokers believing that smoking helps them to relax? In order to optimize the likelihood of changing this belief, the designer can choose from a variety of methods, the so-called behavior change techniques. Different behavior change techniques are useful for different determinants of behavior. Taxonomies of these techniques exists, such as in the Intervention Mapping Protocol, in which the change techniques are grouped per determinant [21]. In our example, the designer would choose a behavior change technique that is appropriate to change attitudes, as the belief to be changed is a perceived advantage of tobacco smoking, thus an attitudinal belief. Several matching behavior change techniques to and for our (attitudinal) change objective are stipulated in the Intervention Mapping Protocol, for example, classical conditioning, self-reevaluation, environmental reevaluation, shifting perspective, arguments, direct experience, elaboration, anticipate regret, and repeated exposure [21]. Thus, we can, for instance, use arguments (i.e., the use of logical facts to stress the advantages of the desired behavior and the disadvantages of the undesired behavior) to persuade the smoker to belief that smoking increases anxiety and tension.

In the last step, the designer can decide on the practical application of this behavior change technique to target the salient beliefs of the smokers. That means, deciding how to deliver the behavior change technique, such as the use of videos, text messages, or games. In tailored dHealth intervention, usually the designer will write health messages that are in line with the matching behavior change techniques.

Once the messages are created, the program needs to be able to tailor the messages to the user's responses on the extent they agree to the salient beliefs. One way to do this is to ask users to rate to what extent this person agrees to a salient belief on a 7-point Likert scale, such as "I think that smoking helps me relax," ranging from 1 (totally disagree) to 7 (totally agree). For allocating the relevant message to the relevant answer, we use certain algorithms. These algorithms can be designed by a health professional who has a basic understanding of programming and then coded by a programmer with the correct syntax in any programming language. Basically, all these algorithms have the if-then-else logic. For instance, "If the response to question 1 is a, then show the user message 1a; If the response to question 1 is b, then show the user message 1b". These questions could be sequenced according to the phases of change as stipulated by the I-Change Model.

Exercise 3: Design your tailored digital health program to enhance smoking cessation

"This is a message from Mylène, the minister of health of The Farmerlands. Tobacco smoking is still a major problem among the adult population Our country has implemented health promotion activities before, but they were not effective. One example is the digital flyer, with the aim to enhance knowledge on smoking among citizens of The Farmerlands; which had almost no effect. My sources informed me that computer-tailoring can be the best strategy to tackle the scourge of tobacco. Therefore, I ask you, a real health promotion expert, to design a tailored dHealth solution! Can't wait for the results and thanks!"

Before proceeding reading this chapter, please design your own tailored dHealth intervention, describing (1) choice of the theoretical model; (2) goal and objectives of the tailored dHealth program; (3) identification of salient beliefs; and (4) design the program content and algorithm.

5. Conclusions

In this chapter, you learned about sociocognitive theories and determinants of behavior to understand the underlying mechanism to promoting health behavior. The critical integration of theories leads to better models to explain behavior, such as the I-Change Model that stipulates three phases of change: awareness, motivation, and action. One pragmatic approach to design dHealth is to (1) choose the theoretical model; (2) formulate intervention goal and objectives of the dHealth program, (3) identify the salient beliefs using qualitative as well as quantitative research, and (4) design the program content and algorithm.

References

[1] Bandura A. Social foundations of thought and action: a social cognitive theory. Englewood Cliffs, NJ: Prentice-Hall; 1986.
[2] Bandura A. The self system in reciprocal determinism. Am Psychol 1978;33(4):344.
[3] Janz NK, Becker MH. The health belief model: a decade later. Health Educ Q 1984;11(1):1—47.
[4] Fishbein M, Ajzen I. Belief, attitude, intention, and behavior: an introduction to theory and research. 1977.
[5] Fishbein M, Triandis HC, Kanfer FH, et al. Factors influencing behavior and behavior change. 2000.
[6] Bandura A. Social foundations of thought and action. 1986. Englewood Cliffs, NJ. 1986.
[7] Ajzen I. The theory of planned behavior. Organ Behav Hum Decis Process 1991;50(2):179—211.
[8] Prochaska JO, Velicer WF. The transtheoretical model of health behavior change. Am J Health Promot September—October 1997;12(1):38—48.
[9] Locke EA, Latham GP. Building a practically useful theory of goal setting and task motivation. A 35-year odyssey. Am Psychol September 2002;57(9):705—17. PubMed PMID: 12237980.
[10] De Vries H. An integrated approach for understanding health behavior; the I-change model as an example. Psychol Behav Sci Int J 2017;2(2):555—85.
[11] Abraham C, Sheeran P. The health belief model. In: Conner M, Norman P, editors. Predicting health behaviour Berkshire. Open University Press; 2005. p. 28—80.
[12] World Health Organisation. Classification of digital health interventions. Geneva: World Health Organization; 2018 (WHO/RHR/18.06). Accessed (19-07-2019): http://apps.who.int/iris/bitstream/handle/10665/260480/WHO-RHR-18.06-eng.pdf;jsessionid=A722F89D4194251E34CAFB302ED855EC?sequence=1.
[13] Cheung KL, Wijnen B, de Vries H. A review of the theoretical basis, effects, and cost effectiveness of online smoking cessation interventions in The Netherlands: a mixed-methods approach. J Med Internet Res 2017;19(6):e230.
[14] Lustria MLA, Noar SM, Cortese J, et al. A meta-analysis of web-delivered tailored health behavior change interventions. J Health Commun 2013;18(9):1039—69.
[15] Taylor GM, Dalili MN, Semwal M, et al. Internet-based interventions for smoking cessation. Cochrane Database Syst Rev 2017;9.

[16] Cheung KL, Schwabe I, Walthouwer M, et al. Effectiveness of a video-versus text-based computer-tailored intervention for obesity prevention after one year: a randomized controlled trial. Int J Environ Res Publ Health 2017;14(10):1275.

[17] Jander A, Crutzen R, Mercken L, et al. Effects of a web-based computer-tailored game to reduce binge drinking among Dutch adolescents: a cluster randomized controlled trial. J Med Internet Res 2016;18(2):e29.

[18] Stanczyk N, Bolman C, van Adrichem M, et al. Comparison of text and video computer-tailored interventions for smoking cessation: randomized controlled trial. J Med Internet Res 2014;16(3):e69.

[19] de Vries H, Brug J. Computer-tailored interventions motivating people to adopt health promoting behaviours: introduction to a new approach. Patient Educ Couns 1999.

[20] Dijkstra A, De Vries H. The development of computer-generated tailored interventions. Patient Educ Couns 1999;36(2):193–203.

[21] Eldredge LKB, Markham CM, Ruiter RA, et al. Planning health promotion programs: an intervention mapping approach. John Wiley & Sons; 2016.

[22] Michie S, Van Stralen MM, West R. The behaviour change wheel: a new method for characterising and designing behaviour change interventions. Implement Sci 2011;6(1):42.

[23] van Gemert-Pijnen JE, Nijland N, van Limburg M, et al. A holistic framework to improve the uptake and impact of eHealth technologies. J Med Internet Res 2011;13(4):e111.

[24] Nierkens V, Stronks K, Van Oel CJ, et al. Beliefs of Turkish and Moroccan immigrants in The Netherlands about smoking cessation: implications for prevention. Health Educ Res 2005;20(6):622–34.

[25] Cheung KL, Evers SM, Hiligsmann M, et al. Understanding the stakeholders' intention to use economic decision-support tools: a cross-sectional study with the tobacco return on investment tool. Health Pol 2016;120(1):46–54.

CHAPTER 9

Illustration of tailored digital health and potential new avenues

Kei Long Cheung, PhD[1], Santiago Hors-Fraile, MSc[2], Hein de Vries, PhD[3]

[1]Department of Health Sciences, Brunel University London, London, United Kingdom; [2]Salumedia Labs, Seville, Spain; [3]Professor in Health Communication at the Department of Health Promotion, CAPHRI Public Health, Maastricht University, Maastricht, The Netherlands

Contents

This chapter illustrates a previously discussed pragmatic methodology to design a tailored smoking cessation digital health program, using the Integrated-Change Model (I-Change model) [1]. Additionally, this chapter will discuss potential new avenues for tailored digital health (dHealth), with a discussion on the use of artificial intelligence (AI) for tailoring.

After reading and conducting an exercise of this chapter, the reader should be better able to

1. Describe how qualitative and quantitative research methods feed into the intervention development
2. Apply and appraise theories and methods to design tailored digital health for smoking cessation
3. Compare and appraise rule-based and data-driven tailoring to elicit behavior change

In the previous chapter, you learned about sociocognitive theories and determinants of behavior, to navigate through health promotion research and practice. The critical integration of theories leads to better models to explain behavior. One integrated model, the Integrated-Change (I-Change) Model, integrates various sociocognitive theories and helps to understand the

Digital Health
ISBN 978-0-12-820077-3
https://doi.org/10.1016/B978-0-12-820077-3.00009-2

159

underlying mechanism of behavior change, stipulating three phases of change: awareness, motivation, and action. One pragmatic approach to apply this model to design behavior change interventions is to start with exploring the beliefs of the target population regards the health behavior. We need to understand which beliefs are important in engaging in a certain health behavior. For this purpose, we need qualitative research, such as via interviews or focus groups. We can use quantitative methods to identify the salient beliefs, the perceptions that statistically explain the behavior. Often, a survey study is conducted to compare healthy and nonhealthy individuals via statistical analyses. The health-promoting intervention should then address these beliefs. To further illustrate the steps to design tailored digital health programs, we will now elaborate on the process using the case of smoking cessation. If you did not yet read the previous chapter and conducted the exercises, we urge you to do it before proceeding this chapter (you haven't done all exercises right? Just do it!).

1. Case of tobacco smoking

As tobacco smoking remains the worldwide leading cause of death, illness, and impoverishment [2], research on this topic to develop and validate tailored dHealth interventions is extremely relevant for society. Thus, the goal of this example intervention will be to increase smoking cessation for motivated and unmotivated smokers. The I-Change Model is relevant to be used, as many evidence-based tailored interventions have used this framework before. Yet, intervention planning protocols, e.g., Intervention Mapping can guide the overall program development and theory selection [3]. The specific objectives can be planned based on the determinants of the chosen model.

When opting to design an intervention that covers any of the behavior change phases (i.e., awareness, motivation, action) [1], it is needed to identify the most relevant beliefs concerning the determinants of each phase. This identification entails looking which items predict the transition from smoking to nonsmoking. Ideally one needs experimental and/or longitudinal results data, but as a second best strategy, one can also compare the means of smokers and nonsmokers on particular items. For example, if smokers and nonsmokers score high on the perception that smoking causes lung cancer, we may not want to put much effort in selecting this variable for developing a message, as both groups are already convinced. Yet, if our results show that smokers are somewhat convinced but still significantly less than nonsmokers, this may be an indication that we do need to incorporate a message on lung cancer for our target group. In this step, we identify which attitudinal beliefs (i.e., which pros and cons) discriminate smokers from nonsmokers, and then select these items for developing the health messages to support smoking cessation. Salient beliefs can also differ per subgroup. For instance, the belief that quitting is good for my unborn baby is relevant for a pregnant smoking woman, but not for those not pregnant. The same strategy holds for identifying messages on how to create social support. Here we assess which people in the social environment of the target group could play a positive or negative role toward supporting a smoker to quit. This could, for example, be the partner, parents, family, friends, colleagues, health professionals, and religious leaders in certain cultures.

Similarly, concerning self-efficacy we need to identify which barriers are relevant for our target group to hinder the realization of the relevant behavior. For instance, for smokers it may be difficult to refrain from smoking when at a party, when with smoking friends, when feeling stressed, when feeling alone, when gaining weight, etc. If these items come up from our analysis of salient beliefs, they will be selected for making self-efficacy messages.

Once we know which determinants are more relevant for our target group, we can select these items to develop the program content (i.e., messages) and the algorithm. At this point, choices need to be made concerning how many intervention interactions (sessions) will be useful, realistic, and feasible. Examples of how these choices are made can be found in so-called study protocols that describe the rationale and make up of computer-tailored programmes (see, e.g., Elfeddali et al., [4]). Making a graphical outline of your program can be helpful to articulate these ideas (see Fig. 9.1 for an example of a computer-tailored program outline [5]).

When it is clear which variables need to be addressed, messages for each variable need to be developed that match a person's score on this variable. For instance, a question could be: "When I quit smoking, my health will improve." A message for those who respond affirmative could be: "*Dear John, you said that quitting smoking will improve your health. There are indeed several advantages for your health, such as reducing your risks for lung cancer and heart problems after a couple of years. But certain health benefits can be noted quite rapidly, such as a much better physical condition. John, you will note this difference immediately when taking a couple of stairs!*" A message for those who respond negatively can be made very easily be reframing some elements of the preceding message. "*Dear John, you said that quitting smoking will not improve*

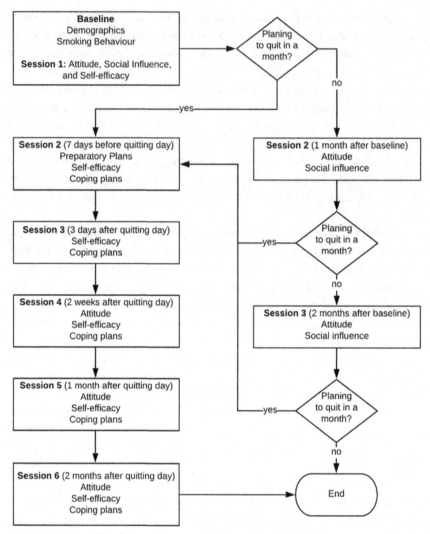

Figure 9.1 Example of a computer-tailored smoking cessation program outline [5].

your health. There are several advantages but they may not become immediately apparent to you, such as reducing your risks for lung cancer and heart problems. These benefits show after a couple of years. But certain health benefits can be noted quite rapidly, such as a much better physical condition. John, you will note this difference immediately when taking a couple of stairs!"

The creation of messages entails both the identification of core constructs and also which behavior change techniques can be used [6,7].

Furthermore, it is important to know whether the computer-tailored intervention does what it is supposed to do. Various types of evaluation exist; one type of evaluation concerns the *user evaluation*. This entails knowing the responses of your users concerning their appreciation of the program to see whether they liked it, which elements were liked most, the level of engagement (e.g., did they use the whole program or only certain elements, did they return to the program after the first session). User evaluation can be done quantitatively (e.g., via a questionnaire after each session with brief questions, or by assessing an overall evaluation score, or via likes), qualitatively via interviews with the target group, or via a thinking aloud procedure in which you let the user respond to the program where you let the user think aloud about his reactions to the program and to monitor their actions. The thinking aloud procedure has become a very popular technique for usability testing for product design and development of various types of interventions in general [8]. Second, effect evaluation is important to see whether your computer-tailored intervention accomplished in realizing the predetermined objectives. One of the core objectives for a smoking cessation program will be to measure whether cessation for which normally 7-day point prevalence and/or prolonged abstinence after 6 or 12 months are chosen as behavior objectives (see Cheung et al. [9]) for a more elaborate discussion on this topic). A recommended effect evaluation strategy is the Randomized Control Trial in which participants are randomly assigned to a condition where they receive the computer-tailored intervention or to a condition in which they do not receive this intervention and receive nothing or usual care. You can then, for instance, hopefully observe that cessation rates in your tailored dHealth condition were 15% versus 5% in the control group, and statistically test whether these results are significant. Several publications of such randomized controlled trials using computer tailoring are available, with the application of the I-Change Model without [10] and with using the Internet [11,12]. Third, you can also perform economic evaluations to assess the costs and effects associated with an intervention and to compare these with the costs and effects of other interventions and/or current practice (see for an explanation, for instance, Smit et al., [13]). The advantage of economic evaluations beyond effect evaluations is that they can demonstrate that although a new intervention may be costlier than current practice, it can be very cost-effective to implement this program, due to economic savings that will result from smoking cessation.

2. New avenues of computer tailoring

Albeit an effective approach to change lifestyle behavior, effect sizes of computer tailoring seem to have reached a ceiling. Additionally, tailored programs have high attrition, which is most likely due to their poor user experience [14]. Therefore, it is important to explore new avenues that potentially take computer tailoring forward [15]. Tailoring the message frame of health messages is a promising avenue. As individuals have different needs for cognition, affect, and autonomy, it is interesting whether tailoring information using different framing could enhance intervention effects [16]. Research can shed light on the effects of different delivery modes and conditions to optimize the different delivery modes in tailoring. For instance, traditional tailoring uses text messages, while as using videos has shown promising effects, outperforming tailoring health messages using text-only [11]. Additionally, the delivery of health messages via smartphones versus personal computers can be explored [17], as well as the delivery via animated videos versus text or actor-based videos. Another avenue, which is the focus in the remainder of this chapter, has high potential to innovate computer tailoring: the use of AI to tailor health messages.

3. Use of artificial intelligence to progress computer tailoring

As of now, like in the example of smoking cessation, tailored dHealth interventions use a rule-based tailoring approach. That means that the program designers select the determinants and develop the rules that specify how the content should be tailored, based on their knowledge of the targeted population, the literature, and health behavior theories. This theory-based, designer-written rules approach may underrepresent the complexity of behavioral change processes, limiting the messages' relevance to the individual. In order for the rule-based algorithm to work, the user needs to feed the program by responding to lengthy questionnaires. The more variables the system needs to tailor on, the lengthier the questionnaire is needed, hampering the user experience.

However, a new paradigm is rising for tailored dHealth programs, using AI: health recommender systems [18]. Instead of the static if-then-else rules in classic rule-based tailored programs, these data-driven algorithms exploit user profiles, the continuous user feedback, and the contents features to empirically tailor the recommendations. That continuous feedback is paramount to make the system learn from user preferences. It is usually shaped as user ratings of the relevance and usefulness of the recommendations sent by the system, although it can be determined in other subtler and intrinsic ways like tracking the user lifestyle (i.e., steps counter) and validating whether the messages are impacting the behavior.

A commonly used principle in this type of systems is collaborative filtering [19], which assumes that similar people benefit from the same things. For example, a message that user "A" (i.e., a young man, living in the city, with no children) considered relevant will also be considered relevant for a similar user "B" (i.e., also a young man, living in the city, with no children), but not to the very different user Z (i.e., old woman, living in the country side, with children). Thus, collaborative filtering needs a similarity function that uses different factors to compute the similarity between users (or items to be recommended). These factors could be the basic demographic features as we used in the example. Nonetheless, any other variable associated to the user profile can be included for the collaborative filtering similarity function. One of the drawbacks these types of systems have is that at the start the system has not yet collected user data for sufficient collaborative filtering (also known as the cold-start problem [3]) and would therefore deliver/tailor messages at random. Yet, when the system gathers more data, and therefore "warms-up," it would start tailoring more accurate recommendations.

Among the multiple other types of AI algorithms that can be applied [18], we argue that algorithms based on easy-to-explain principles like collaborative filtering are convenient and useful for the purposes of tailoring behavioral change interventions. They offer a good trade-off between the provided performance, the capacity of behavioral change experts to act on them and be involved in their design, and the cast trust in users. Users who understand how a recommendation (i.e., the tailored health message) was generated may accept and embrace the intervention, following its recommendations [20].

Exercise. Tailoring using artificial intelligence

You have decided to implement a tailored behavioral change intervention for smoking cessation in The Farmlands using AI under the principle of collaborative filtering.

You choose to design a similarity function based on the previously stated preferences of each user. To illustrate this with a simple example, let us assume

---

Continued

Exercise. Tailoring using artificial intelligence—cont'd

we only have three smokers, and two of them have received the same 10 behavioral change messages (M1–M10). We have detected that user A benefits from messages M1, M3, M5, M7, and M9 and user B benefits from M2, M4, M6, M8, and M10. Now, user C has received messages from M1 to M8, but only benefited from M1, M5, and M7. If the system can only send one extra message to user C (M9, or M10), based on the similarity function we described, which one would be sent by the AI intervention?

Answer: Despite not having benefited from the messages exactly in the same way as user A, who also benefited from M3, the AI system would consider user C more similar to user A than to user B as there are more coinciding messages that benefited in common. Therefore, if only one additional message can be sent to user C, it would be M9 instead of M10 because M9 was also useful for user A.

The AI-based approach has the potential to tap into many variables and patterns of the change process, deriving the tailoring rules from the collective-intelligence data of the individual, as well as of the group. The ability to incorporate many variables and complex patterns may enhance the relevance of the tailored health messages compared to the traditional rule-based interventions. In addition, the system has the capability to learn from the changing needs of the users and adapt the recommendations to these changes. Furthermore, relying on AI may also remove the burden of lengthy questionnaires in traditional rule-based tailoring, which could enhance the user experience as well [14]. Despite not being widespread in health promotion interventions, AI has the potential to improve the impact of tailored dHealth. It is argued that combining AI with a theory-steered approach has potential to increase users' perceptions of the efficiency, effectiveness, trustworthiness, and enjoyment. It is expected that this in turn enhances the usage of these programs. As this stream of research is relatively unexplored, the best way forward in AI tailoring is still unknown. Yet, tailoring through AI has high potential to be the next frontier for (tailored) dHealth programs.

A successful example of the AI tailoring approach was illustrated within the European H2020 SmokeFreeBrain project [21]. In this project, a multidisciplinary group of experts—including healthcare professionals, behavioral scientists, and software engineers—developed a mobile intervention to support smoking cessation using AI to tailor health messages [22]. The AI algorithm used the stages of change of the Transtheoretical model (i.e., the readiness to quit smoking was assessed) to determine how many messages will be sent to the users, and used explicit user data (via a brief questionnaire) and implicit user data (via in-app usage analytics) as inputs for the similarity function [23]. A second version based on this mobile intervention was developed during the project, using the I-Change model as theoretical framework instead [24]. For this new version, more lifestyle factors and determinants of the I-Change model were embedded in the AI algorithm design. This system was tested in clinical trials in Spain and Taiwan and was shown to help patients increase their abstinence rates compared to the control groups [25–27].

4. Conclusions

Sociocognitive theories, such as the I-Change Model, are useful when planning interventions. There are elaborated planning models, but they can be deemed highly complex and resources may be limited. The previous chapter and this chapter outlined and illustrated one pragmatic approach to design digital health, basically using a four-step approach. This approach helps to shape evidence-based computer-tailored interventions. Despite the effectiveness of computer tailoring, effect sizes of computer tailoring seem to have reached a ceiling. The exploration how to progress in tailored dHealth is therefore important. Streams of research could shed light on the use of different delivery modes, and message framing in tailoring. One approach that is especially interesting is the use of AI to innovate the algorithms of tailoring, moving from rule-based to data-driven tailored dHealth. Although promising examples like the system created and validated in the SmokeFreeBrain project—now commercialized as DigiQuit (www.digiquit.com)—are showing the benefits of tailoring interventions using AI, they have potential limitations. One has to be thoughtful regards the design and development of these systems, as a single incorrect recommendation can be harmful for the user and can make the user lose trust in the program [20]. A potential notable limitation of tailoring using AI is that health promoters do not control the final recommendation to the user as the user data are constantly changing which alter the recommendation of health messages [28]. For instance, if there are malicious users providing intentionally incorrect feedback to the system, the recommended health messages may not align to the user needs. This is in contrast with the traditional rule-based tailoring where the health promoter has full control on the provision of health messages based on the specific user input. Using validated solutions or working with a multidisciplinary team of experts to be able to understand clinical, psychological, and AI areas is of utmost importance to achieve safe and satisfactory results combining AI and tailoring to induce behavioral changes.

References

[1] De Vries H. An integrated approach for understanding health behavior; the I-change model as an example. Psychol Behav Sci Int J 2017;2(2):555–85.
[2] World Health Organisation. Leading cause of death, illness and impoverishment. Tobacco; n.d. [Accessed 20 July 2019]: https://www.who.int/news-room/fact-sheets/detail/tobacco.
[3] Methods and metrics for cold-start recommendations. In: Schein AI, Popescul A, Ungar LH, et al., editors. Proceedings of the 25th annual international ACM SIGIR conference on research and development in information retrieval. ACM; 2002.
[4] Elfeddali I, Bolman C, De Vries H. SQ4U—a computer tailored smoking relapse prevention program incorporating planning strategy assignments and multiple feedback time points after the quit-attempt: development and design protocol. Contemp Clin Trials 2012;33(1):151–8.
[5] Stanczyk NE, Bolman C, Muris JW, et al. Study protocol of a Dutch smoking cessation e-health program. BMC Publ Health 2011;11(1):847.
[6] Eldredge LKB, Markham CM, Ruiter RA, et al. Planning health promotion programs: an intervention mapping approach. John Wiley & Sons; 2016.
[7] Webb T, Joseph J, Yardley L, et al. Using the internet to promote health behavior change: a systematic review and meta-analysis of the impact of theoretical basis, use of behavior change techniques, and mode of delivery on efficacy. J Med Internet Res 2010;12(1):e4.
[8] Jaspers MW. A comparison of usability methods for testing interactive health technologies: methodological aspects and empirical evidence. Int J Med Inf 2009;78(5):340–53.
[9] Cheung KL, de Ruijter D, Hiligsmann M, et al. Exploring consensus on how to measure smoking cessation. A Delphi study. BMC Public Health 2017;17(1):890.
[10] Dijkstra A, De Vries H. The development of computer-generated tailored interventions. Patient Educ & Couns 1999;36(2):193–203.
[11] Stanczyk N, Bolman C, van Adrichem M, et al. Comparison of text and video computer-tailored interventions for smoking cessation: randomized controlled trial. J Med Internet Res 2014;16(3):e69.
[12] Te Poel F, Bolman C, Reubsaet A, et al. Efficacy of a single computer-tailored e-mail for smoking cessation: results after 6 months. Health Educ Res 2009;24(6):930–40.
[13] Smit ES, Evers SM, de Vries H, et al. Cost-effectiveness and cost-utility of internet-based computer tailoring for smoking cessation. J Med Internet Res 2013;15(3):e57.
[14] Cheung KL, Durusu D, Sui X, et al. How recommender systems could support and enhance computer-tailored digital health programs: a scoping review. Digit Health 2019;5. 2055207618824727.
[15] Smit ES, Linn AJ, van Weert JC. Taking online computer-tailoring forward. Health 2013;17(1):25–31.
[16] Altendorf MB, van Weert JC, Hoving C, et al. Should or could? Testing the use of autonomy-supportive language and the provision of choice in online computer-tailored alcohol reduction communication. Digit Health 2019;5. 2055207619832767.

[17] Quiñonez SG, Walthouwer MJL, Schulz DN, et al. mHealth or eHealth? Efficacy, use, and appreciation of a web-based computer-tailored physical activity intervention for Dutch adults: a randomized controlled trial. J Med Internet Res 2016;18(11):e278.

[18] Sadasivam RS, Cutrona SL, Kinney RL, et al. Collective-intelligence recommender systems: advancing computer tailoring for health behavior change into the 21st century. J Med Internet Res 2016;18(3):e42.

[19] PERSPeCT: collaborative filtering for tailored health communications. In: Adams RJ, Sadasivam RS, Balakrishnan K, et al., editors. Proceedings of the 8th ACM Conference on Recommender Systems. ACM; 2014.

[20] Towards health (aware) recommender systems. In: Schäfer H, Hors-Fraile S, Karumur RP, et al., editors. Proceedings of the 2017 international conference on digital health. ACM; 2017.

[21] Multimodal e-health services for smoking cessation and public health: the smoke-freebrain project approach. In: Bamidis PD, Paraskevopoulos E, Konstantinidis E, et al., editors. MEDINFO 2017: Precision healthcare through informatics: proceedings of the 16th World Congress on Medical and Health Informatics. IOS Press; 2018.

[22] Hors-Fraile S, Benjumea FJN, Hernández LC, et al. Design of two combined health recommender systems for tailoring messages in a smoking cessation app. 2016. arXiv preprint arXiv:160807192.

[23] Hors-Fraile S, Schneider F, Fernandez-Luque L, et al. Tailoring motivational health messages for smoking cessation using an mHealth recommender system integrated with an electronic health record: a study protocol. BMC Public Health 2018;18(1):698.

[24] Hors-Fraile S, Malwade S, Spachos D, et al. A recommender system to quit smoking with mobile motivational messages: study protocol for a randomized controlled trial. Trials 2018;19(1):618.

[25] Coupling neuroscience smoking cessation interventions with social media and mobile devices. Front Hum Neurosci. In: Hors-Fraile S, Civit A, Nuñez Benjumea F, et al., editors. Conference Abstract: SAN2016 Meeting; 2016.

[26] Jódar-Sánchez F, Hernández LC, Núñez-Benjumea FJ, et al. Using the social-local-mobile app for smoking cessation in the SmokeFreeBrain project: protocol for a randomized controlled trial. JMIR Res Protoc 2018;7(12):e12464.

[27] Carrasco-Hernández Laura, Jódar-Sánchez Francisco, Núñez-Benjumea Francisco, Moreno Conde Francisco, Civit-Balcells Antón, Mesa González Marco, et al. A mobile health solution complementing psychopharmacology-supported smoking cessation: randomized controlled trial. JMIR mHealth and uHealth 2020;8(4):e17530. https://doi.org/10.2196/17530. https://mhealth.jmir.org/2020/4/e17530/; 2020.

[28] Hors-Fraile S, Malwade S, Luna-Perejon F, Amaya C, Civit A, Schneider F, et al. Opening the black box: explaining the process to base a health recommender system on the I-change behavioral change model. IEEE Access 2019;7:176525−40.

CHAPTER 10

Sustainability of mHealth solutions for healthcare system strengthening

Mohamed-Amine Choukou, PhD
Department of Occupational Therapy, College of Rehabilitation Sciences, University of Manitoba, Winnipeg, MB, Canada

Contents

This chapter discusses the mHealth development process, the role of mHealth in healthcare strengthening, and the determinants of mHealth solutions sustainability.

1. Introduction

In recent years, interest in mHealth technologies has grown, partly due to low budgets allocated toward the health sector, health costs, and the burden of patients, caregivers, and care providers; therefore, the goal of mHealth solutions is to strengthen the capacity of the healthcare system without sacrificing the quality of care. Delivering low quality healthcare services increases the burden of illness and healthcare costs globally [1]. A major contributor to the healthcare system fragmentation is the traditional

Digital Health
ISBN 978-0-12-820077-3
https://doi.org/10.1016/B978-0-12-820077-3.00010-9

philosophy of care, which considers the patient—care provider interaction as a basic commercial customer-service provider relationship, in which the client has to navigate the healthcare system in order to understand where the primary care is provided, the whole process of care and then wait for appointments to be scheduled. This process is more likely to function correctly as a "standard" service delivery approach, but it can rapidly turn into a nightmare when it comes to inquiring about healthcare services. Human health is much more important than any other daily life need. Health is vital to human existence and caring for humans is much more complex than delivering anticipated and well-established solutions. The patient—care provider relationship is a central part of healthcare and is based on high expectations in terms of exchanging information during care encounters. Patient information evolves over time and in traditional healthcare settings, the healthcare provider cannot access the information with a simple, yet logistically challenging monitoring approach. Therefore, access to information is indeed a key to prevention and early intervention.

Nowadays, technologies offer several ways for real-time information sharing and virtual dialogue. From a chronological point of view, healthcare could be considered one of the last ecosystems to adhere to the concept of digital data sharing and electronic messaging. mHealth is among the newest e-concepts that emerged in the last decades following the generalization of mobile devices and wearable technologies. mHealth has been defined by the World Health Organization as a "medical and public health practice supported by mobile devices, such as mobile phones, patient monitoring devices, personal digital assistants, and other wireless devices" [2]. Endless mobile applications (apps) are available on the market but news about long-term adoption or real changes in the healthcare system are rather rare. It is even surprising to know the extent to which a program is sustained over time is one of the least reported outcomes across health literature [3].

The ultimate goal of this chapter is to discuss the facilitators and barriers to developing sustainable mHealth solutions and to draw a portrait of the gaps in the implementation and adoption of mHealth solutions in practice. This chapter will elucidate the way development endeavors can go beyond proof-of-concepts and pilot implementations and translate into sustainable innovations. In particular, this chapter has three objectives:

- Demystify the concept of mHealth and its implication on patient-centeredness and the healthcare system as a whole,
- Take a critical look at enablers and barriers to using mHealth solutions to strengthen the existing healthcare systems, and
- Draw a portrait of mHealth solutions sustainability determinants.

2. mHealth solutions: disruptive technologies as a remedy for a system under strain

Increases in health operating and capital costs and in care providers and families' burden are indisputably unbearable for the community. To cope with these consistent increases, politicians, academics, and industry leaders worldwide are looking for innovative approaches to address these threatening issues and increase access to quality care and decrease health inequities. For the past 2 decades, there has been a consensus among stakeholders that mHealth is a good line of sight to follow. mHealth is indeed perceived as a way to enable cost reduction and deliver quality outcomes for patients and families for less. It is not surprising that mHealth is the heart of innovative solutions and that several private companies around the globe are competing to lead in this industry. Current mHealth solutions could ultimately redefine healthcare delivery. Apps are being developed globally to target all the healthcare system stakeholders including the patients, family caregivers, care providers, and the management systems. Table 10.1 presents examples of main mHealth applications for patients and healthcare providers.

mHealth solutions are created essentially to increase the efficacy of patient—care provider interactions and reduce the frequency of in-person interactions [13]. According to Bavafa et al. (2018), e-visits provide a new (and usually low cost) ways for patients to connect with their healthcare providers. They are expected to decrease office and telephone visits [13]. However, according to Bavafa et al. (2018), there is no guarantee that the right types and amounts of data between patients and providers will be shared during an e-visit. Electronic communication can also be perceived by e-visit adopters as a way of circumventing the usual practice (e.g., gatekeepers, office staff, and nurses). If this is the case, a lot of contact with the doctor requires the doctor to see the patient in the office or to have a telephone conversation with the patient, which does not reduce the number of office and telephone calls but rather raises them [13]. One of the disruptive innovations that mHealth suggests is the virtual care which replaces or reduces face-to-face interactions between patients and care providers without altering the overall quality of care. In fact, virtual care aims at reducing wait times and the burden related to arranging—and getting to a face-to-face appointment. Interestingly, mHealth apps are suggesting endless ways for patients to benefit from artificial intelligence to increase their literacy and inform their own decisions independently from the care providers whenever possible, hence, maximizing the focus during virtual care interactions or face-to-face encounters.

Table 10.1 Example of mHealth examples.

mHealth application	User	Use case	Example of use
Behavior change communication	Patient	Text messaging for chronic disease management.	• Monitoring self-management behaviors between clinic visits [4]. • Monitoring medication adherence in chronic disease [5].
Service delivery	Patient	Provide follow-up visits or electronic decision support.	• Monitoring for—and—early detection of deterioration in heart failure [6]. • Provide earlier diagnosis through remote transmission and interpretation of ECG, leading to more accurate triage and shorter door-to-balloon time in myocardial infarction. • Monitoring arrhythmias in real-time using automated detection algorithms, and implantable device status checks.
Financial transactions and incentives	Patient	Provide incentives to clients to use particular areas of health services.	Institutional deliveries or vaccines [7]. Electronic food/medicine vouchers are distributed as an incentive to mothers visiting immunization programs in low socioeconomic area and completing the required vaccines for their infants on time.

Table 10.1 Example of mHealth examples.—cont'd

mHealth application	User	Use case	Example of use
Information and data collection systems	Healthcare provider	Use electronic methods and dispatch the information to various levels of healthcare to increase the speed of information processing to benefit both the patient and family caregiver. Information systems help keep health information connected so healthcare providers can access the information when needed and share appropriate information on the go.	mHDC is an mHealth data collector app used to collect and report indicators for assessment of cardiometabolic risk. mHDC is used in population surveys that contain such data as body mass index, health-related issues, and health habit indicators [8].
Logistics and supply management	Healthcare provider	Pharmacists and other service providers send a structured text message periodically to their district coordinator to report stock levels at the end of the week. The supervisor can then take necessary actions to monitor the flow of the commodities and their availability in the market.	In Tanzania, more than 130 clinics are using the SMS for Life mHealth supply chain system to prevent stockouts of essential malaria drugs [9].
Workforce development	Healthcare provider	Important aspect from a professional and systemic point of view as soft skills are increasingly needed and will indirectly enhance the patient's outcome in the long run [10,11].	eMOCHA is a platform that allows frontline health workers in rural Uganda to select streaming video content as part of continuing education [12].

3. Digital twin technology as an example of technology-driven healthcare system change

Digital twin technology is enabling powerful patient—care provider interactions by giving care providers the tools and channels to get closer to the patient more than ever before. Indeed, it offers a digital replica for the physical object or face-to-face service they represent in the healthcare industry, providing remote evaluation, ubiquitous monitoring, and powerful prediction tools. While digital twinning is still in its early stages, its first achievements promise more holistic and realistic portraits of the condition of each unique patient. Applied to telerehabilitation [14], digitizing the human body enables a new level of the patient—care provider experience. Being evidence-based and built by experts for experts, emerging mHealth technologies are more advanced, more accurate, and faster communicators than humans. Whether they are deployed in institution or remotely, digital twin technologies are key to healthcare cost reduction and health system strengthening. The predictable future of human body digitization is a combination of the replica of the patient physical body and/or environment with virtual reality technologies to reach new heights of realism in the virtual simulation experience. Digital twin technologies look virtual at first glance, but they might be humanizing the virtual experience even more than expected. This is an asset to be adopted and sustained in the healthcare system.

Although it has great potential to strengthen patient—care provider interactions, using digital twinning appropriately for healthcare purposes is challenging considering the weaknesses of the traditional health system in terms of information availability and sources, including, but not limited to, incomplete patient information, disjointed flow of information, and various disconnected sources of patient data. Digital twin technology has the capacity to meet patient's needs if the information needed is clear, complete, and available, thus providing a comprehensive approach to characterize each unique patient and trigger the right processes of care. Therefore, one can consider that digital twinning is conditional to information being available, digitized and gathered in the same patient portfolio, a next step in patient-centric care. A digitally empowered patient would have a better care experience, as reported in chronic disease patients (e.g., diabetes self-management system [15]), feel in control of their condition hand in hand with the care providers, and feel ready to make a decision with the care providers relying on the updated data available in their portfolio. Patients informed about their condition and who have access to digital information about their health state would be more engaged and proactive in asking for help and orientation, and more involved in shared decision-making processes [16,17].

4. mHealth solutions sustainability

Theoretically, it is clear that mHealth solutions are key to health system strengthening and it is a reality that mHealth will rapidly evolve over time and adapt to continuing users' need. Particularly, mHealth solutions are technologies that are introduced to a vital and high demanding field that has been under strain for a long time. In the healthcare system, there are several areas of contention from different stakeholders and it is difficult for mHealth solutions to find frameworks that would allow them to spread faster to their full potential and remain permanently in the healthcare system. mHealth is clearly a cultural shift of the healthcare system from a traditional ("paternalistic") model of care to an equal level interaction model between patients and care providers based on consistently sharing information throughout the mHealth app and following shared decision-making processes. However, this cultural shift faces many barriers to its implementation. There is mainly a lack of clear legal and regulatory frameworks for the implementation and adoption of mHealth solutions whether they are apps, software, or programs [18—24]. Furthermore, there is reluctance among care providers ([25—27]) whose work may be constrained to focus on filling gaps temporarily to produce better short-term outcomes, more likely because their actions are in tandem with short-term healthcare policies. A survey conducted in 2005 has shown that physicians are reluctant to the use of telemedicine and expect the same reluctance from their patients [28]. For example, according to Gaggioli et al. (2005), the majority of the surveyed doctors (361 Italian physicians) estimated that less than 25% of their patients would be advantaged by telemedicine and that less than 25% of their patients would agree to be involved in telemedicine [28]. mHealth is many steps ahead of traditional healthcare approaches, but there are many more steps to do before bringing technology-oriented behavioral changes to both patients and care providers to induce healthcare system strengthening. In other words, healthcare strengthening relies on making the system function better and sustainable for everyone. The democratization of mHealth solutions is a complex process that faces barriers from different points of view. Fig. 10.1 suggests a way of summarizing the characteristics of sustainable mHealth technologies from economic, environmental and social points of view. According to this framework, sustainable mHealth solutions should be economically viable, environmentally bearable, and socially equitable.

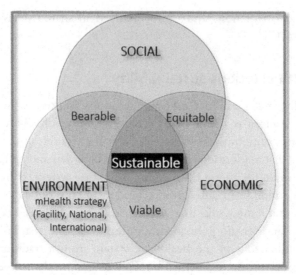

Figure 10.1 Triple bottom line framework adapted to mHealth technology sustainability.

To maintain or sustain mHealth solutions, the technology-based interventions may need to change over time to adapt to the conditions surrounding its use including, but not limited to, healthcare needs, resources and structures available, environmental characteristics, economical changes, and social acceptance [29,30].

4.1 Economically viable mHealth strategy

mHealth solutions developments need to take into account the fact that appropriate sources of financing are allocated. *Who will pay for mHealth programs?* Indeed, developing and maintaining technology can be costly, requiring investments in devices and training for users. In healthcare systems struggling to control costs, financing mHealth could be considered one of the most important barriers to scaling and implementation. Practically, convincing decision-makers of the potential of mHealth technologies, even in resource constraint situations, will be key to ensuring mHealth solutions financing. An alternative financing model could be the use of open-source platforms that provides lower cost alternatives in underfunded healthcare systems (e.g., open-source short message service-based tool to monitor malaria in remote areas of Uganda [31]).

4.2 Environmentally bearable mHealth strategy

mHealth solutions must be adapted to their environment because every healthcare ecosystem is different. Whether it is at facility level or among regional collaboration initiatives, education is key. Patients, care providers, and administrators need to feel confident about the security and respect of privacy and understand how mHealth apps and programs work as day-to-day tools, so that they contribute successfully to their path to care and voluntarily take advantages of the benefits offered by the mHealth program that they are involved in. The second key is mHealth intervention inter-operability. mHealth apps and platforms will need to interact in a coherent way with each other within healthcare systems to insure a high level of information integration and communication. mHealth solutions will also need to operate across multiple organizations, such as insurances or workers' compensation boards to facilitate the convergence of health information in all aspects of healthcare.

4.3 Socially equitable mHealth strategy

Implementation and confidence in mHealth are influenced by the privacy, safety, and security standards set within a given environment to define responsibilities and malpractice liability. Few regulations, such as the Health Insurance Portability and Accountability Act (HIPAA) of 1996, are available. However, standards need to be enhanced and continuously adapted to different context of use in a way they can be legislated and clearly adopted in the community. It is "likely that malpractice lawyers will take advantage of emerging" [32] mHealth app as a new opportunity for litigation. In the presence of strong standards in place, users will become aware about the legal framework and will be able to identify optimal use and potential benefits. Patients and clinicians will then have the resources to make a meaningful contribution to improving the quality of treatment as part of a socially responsible model of healthcare (for example, setting layers of access to information to make it impossible for mHealth app users to access information they are not meant to have access to).

5. mHealth solutions: path to sustainability

As a technological solution, mHealth platforms have to go through challenges before they are successfully implemented. Challenges are manifold and affect all the stages of the development. Fig. 10.2 shows the

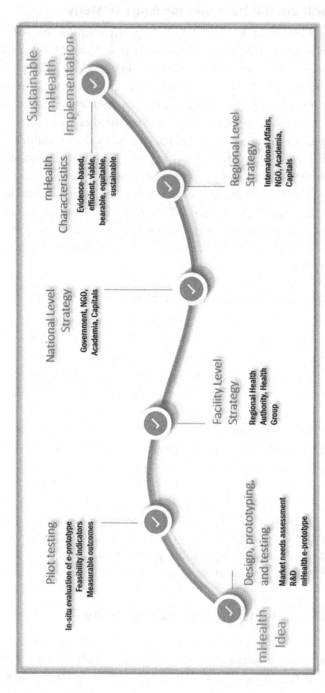

Figure 10.2 Six step journey from ideation to sustainable implementation. *(Based on Scott RE, Mars M. Principles and framework for eHealth strategy development. J Med Internet Res 2013;15(7):e155—e155.)*

sustainability path to be followed as mHealth solutions are established. Sustainable development starts from the design of a pragmatic mHealth idea that grows following the phases of informed development, scientific evaluation, and strategic integration until the level of implementing a sustainable mHealth solution in the community is reached (Fig. 10.2).

Challenges to developing mHealth solutions are manifold and could be introduced from the stakeholders' perspectives. Although it is not an exhaustive examination of the determinants of mHealth solutions sustainability available in the health literature, Tables 10.2 and 10.3 illustrate a summary of the major determinants as they relate to stakeholders: barriers and enablers, respectively. Tables 10.2 and 10.3 would guide the developers and decision-makers to clearly situate their needs, choose the right processes, and set appropriate criteria for promoting the implementation of the most relevant mHealth solutions representing a high potential to strengthen the healthcare system.

6. Conclusion and future directions

The goal of this chapter was to draw a portrait of the determinants of mHealth solutions sustainability to raise the awareness about mHealth sustainability as a key for healthcare system strengthening. In other words, this chapter highlights the challenges of developing mHealth solutions that go beyond the prototype stage and are deployed in care delivery programs. The path from ideation to technology implementation has been demystified and followed by a review of the enablers and barriers to the sustainability of mHealth applications from the point of view of the patient and healthcare provider, as well as technology and leadership. While mHealth solutions seem adapted to their context of use and the users, all the viewpoints mentioned above will require improvement to tackle current challenges such as connectivity, infrastructure, accessibility of the technology, user's acceptance, user's literacy, data privacy, and, above all, effectiveness of the intervention and efficacy of the patient—provider communication. The key and the future of mHealth applications sustainability is interoperability between platforms and across multiple organizations, such as insurances or workers' compensation boards to facilitate the convergence of health information in all aspects of healthcare.

Despite the fact that there is a plethora of mHealth solutions, there are no specifications or regulations guiding the design of mHealth apps and for users to accept and embrace the technology. There is no single way to trust

Table 10.2 Portrait of the barriers to mHealth solutions sustainability.

	Challenges to mHealth sustainability					
	Technology and health literacy	Access	Satisfaction	Support	Tailoring to patients	Clinician–patient communication and collaboration
Patient	Patients lack confidence in internet-based communication and prefer using phone calls. Low health literacy and low formal education [34].	Areas may be outside the coverage range of the mobile telephone network—lack of mobile devices or a single mobile device shared among family members, affordability of airtime, keeping the mobile phone charged in rural areas that are far from the grid—unstable network connectivity [35].	Users who are dissatisfied with a system are likely to develop a negative attitude that may later discontinue use of the system [35].	Service quality and follow-up with users is challenged by the technical updates and trouble shooting.	Lack of customization to patient preferences and needs—content not engaging or relevant—timing of patient–provider interactions—decisions of content and frequency of interventions—patients not incorporated into the design needs—no analysis on impact with comorbidities [34].	Technical difficulties are barriers to telehomecare programs, as they might impact the quality of telehomecare communication, which could in turn be perceived as unhelpful in establishing effective clinician–patient communication and collaboration [36].
	Perception of effectiveness					Interprofessional communication and collaboration
Care provider	Lack of orientation to mHealth programs and training to perform telehomecare interventions, lack of understanding on mHealth use to improve intended clinical outcomes [36]. Patients report more positive experiences with mHealth than do clinicians [36]. Clinicians perceive that patients' reliance on clinicians' mHealth monitoring introduced a "paradoxical threat to patients' independence and self-reliance" [36].					Clinicians need to develop shared goals for patient outcomes and are frustrated from a lack of interoperability between mHealth services providers and current patient information systems in primary care [36].

technology	adoption Low perceived usefulness and perceived ease of use of mHealth solutions reduces the intention of users to use them in the future [37].	mHealth systems that are designed and deployed without provisions and plans for future interoperability with other systems may later be abandoned because of inefficiencies, time, and costs associated with data conversion from one system to another [35].	Failure of mHealth projects attributed to the lack of sufficient expertise in the design, development, implementation, maintenance, and capacity to provide technical support [35]. "mHealth systems that do not pay for themselves will not be sustainable" [34].	Technological solutions with features that do not match the tasks will not be utilized or may be abandoned by users [35].	Complex mHealth systems that are not easy to learn, have poor production quality, low levels of accuracy, longer response times, challenges in availability and reliability, and poor security implementation will negatively impact the use of the system, leading to low or nonuse of the system [35].	mHealth systems require upgrades, updates, and modification to deal with security requirements, processing needs, and collaboration and interoperability requirements when different systems are interfaced. When systems are built using proprietary software, royalties must be paid to obtain licenses for higher versions of the software [35].	considerations Explicit legal regulation is needed. Both need to be taught in academia and in the community. Security issues and legal accountability might represent a significant barrier to implementing mHealth solutions [38]. Provider's access to patient logs [34]. Example: "clinicians might be held accountable for missing signs and symptoms in telehomecare data that were indicative of a patient in crisis" [36].
Management	Resources and infrastructure Absence of wireless networks and the mobile phones is a major barrier to accessing mHealth platforms.	Ownership Governance structure is not clearly defining the owner of the mHealth solutions and the resources to implement, maintain, and enhance the technology.			Net benefits An mHealth solution that does not generate satisfactory benefits will be discontinued inevitably.	Provider Labor- and time-intensive for providers—increased workload [36] and low integration into provider work flow [34].	Process and culture mHealth could be perceived as a "threat to clinicians' identity."

Table 10.3 Portrait of the enablers of mHealth solutions sustainability.

	Enablers of mHealth sustainability					
	Technology and health literacy	Access	Satisfaction	Support	Tailoring to patients	Clinician–patient communication and collaboration
Patient	Patients could use computers and the internet to access and monitor the information.	Coverage range of the mobile telephone network could be expanded—mobile devices are becoming very smart and less expensive so that everyone can has his own smart phone.	Users who are dissatisfied with a system could share their feedback online to resolve their own issues and enhance the performance of the mHealth solution per se.	Service quality could be enhanced by reliable attributes, quick responsiveness to the users' needs, and fast follow-up service [35].	mHealth solutions could be customized to patient preferences and needs—content could be more engaging and relevant—timing of patient–provider interactions could be lowered—alerting systems could trigger almost instantaneous responses—frequency of interventions could be set up by the users—patients engaged into the design needs—more integrative mHealth solutions to address comorbidities adequately [34].	Telehomecare programs (i) drive increased reassurance thanks to improved access to care providers; (ii) enhance providers' understanding of patients' condition as well as health literacy [36]; and (iii) sustain adherence in telehomecare interventions thanks to timely feedback according to mHealth data flow [36].
Care provider	Perception of effectiveness					Interprofessional communication and collaboration
	More orientation to mHealth programs and training to perform telehomecare interventions is needed. More clear information about mHealth use would improve intended clinical outcomes. Patients report more positive experiences with mHealth than do clinicians [36]. Clinician–patient experience could be boosted by a telemonitoring process focused on shared goals and set up in advance following a discussion on the duties and rights to increase patients' independence and self-reliance.					Current patient information systems in primary care could be connected to other health services through mHealth platforms to increase interoperability between mHealth services providers and current patient information systems.

Design/technology	Acceptance/adoption	Interoperability	Scalability	Relevance	Quality	Sustainability	Ethics and legal considerations
	High-perceived usefulness and perceived ease of use of mHealth solutions increases the intention of users to use them in the future [37].	mHealth systems could be designed and deployed with provisions and plans for future interoperability with other systems to increase efficiencies, reduce time, and costs associated with data conversion from one system to another.	Success of mHealth projects could be guaranteed by the availability of sufficient expertise in the design, development, implementation, maintenance, and capacity to provide technical support.	Developers should consider the tasks as well as the environment and the locality's cultural dynamics [35].	High-level mHealth solutions are complete, reliable, and safe, easy to learn, and easy to use [35].	Like any other digital tool, optimization could be implemented in the mHealth systems upgrades. Updates and modification to deal with security requirements could be done less frequently. Processing needs and collaboration and interoperability requirements when different systems are interfaced could be provided online or by phone. Systems could be adopted by the healthcare systems and made available to users for free or with reasonable and stable cost [35].	Ethical considerations could be addressed at early stages of development. Explicit legal regulation could be addressed. Both could to be taught in academia and in the community and introduced to mass media. Security issues and legal accountability could be addressed as part of a social innovation plan.

Continued

Table 10.3 Portrait of the enablers of mHealth solutions sustainability.—cont'd

	Enablers of mHealth sustainability				
Management	Resources and infrastructure	Ownership	Net benefits	Provider	Process and culture
	Availability of wireless networks and mobile phones are facilitators to alleviating the deficits in physical networks: geographic dispersion, high costs of services, percentages of people living in out-of-grid areas [38].	The right of possessing, managing, controlling, and directing the use of the technology. The ownership of a technology defines and influences the environment where the technology is used, the governance of the technology, the resources—manpower and financing available to support the implementation and use of the technology, the strategy and planning for growth of the system [35].	Net benefits accrued from using an mHealth solution will drive further investment into it and hence contribute to its sustainability [35].	Complete turnkey mHealth solutions less challenging to implement and integrate into provider work flow.	mHealth changes traditional care and the patient–care providers interaction because of (i) installing devices, orienting and training the users, and troubleshooting, (ii) tasks related to telehomecare data monitoring, and (iii) increased communication and interaction with patients [36].

the solutions available and to classify the most relevant ones according to the scope of use. There is also not a single official scientific certification for the mHealth solutions to be approved. Maintaining a registry of accredited mHealth solutions and issuing an official scientific approval seal could therefore help decision-makers in implementing a solution. Developers and researchers are encouraged to work together under inter-disciplinary umbrellas to develop and adopt a framework for assessing the usability and efficacy of mHealth solutions and to serve as guidelines to be adopted early on in the new developments. Such a policy will alleviate perplexity and concerns about the validity of mHealth solutions by facilitating the cycle of implementing and retaining accredited mHealth solutions in the healthcare system for the good of clinicians, families, and the healthcare system as a whole.

References

[1] OECD, Organization WH, Group WB. Delivering quality health services: a global imperative. 2018.
[2] WHO, mHealth. New horizons for health through mobile technologies: second global survey on eHealth. In: Global Observatory for eHealth series, vol. 3; 2011 [Geneva].
[3] Gaglio B, Shoup JA, Glasgow RE. The RE-AIM framework: a systematic review of use over time. Am J Public Health 2013;103(6):e38−46.
[4] Fischer HH, et al. Care by cell phone: text messaging for chronic disease management. Am J Manag Care 2012;18(2):e42−7.
[5] Thakkar J, et al. Mobile telephone text messaging for medication adherence in chronic disease: a meta-analysis. JAMA Intern Med 2016;6(3):340−9.
[6] Honeyman E, Ding H, Varnfield M, Karunanithi M. Mobile health applications in cardiac care. Intervent Cardiol 2014;6(2):227−40.
[7] Chandir S KA, Hussain H, Usman HR, Khowaja S, Halsey NA, Omer SB. Effect of food coupon incentives on timely completion of DTP immunization series in children from a low-income area in Karachi, Pakistan: a longitudinal intervention study. Vaccine 2010;28(19):3473−8.
[8] Shishido HY, Alves da Cruz de Andrade R, Eler GJ. mHealth data collector: an application to collect and report indicators for assessment of cardiometabolic risk. Stud Health Technol Inform 2014;201:425−32.
[9] Githinji S, et al. Reducing stock-outs of life saving malaria commodities using mobile phone text-messaging: SMS for life study in Kenya. PLoS One 2013;8(1):1−8.
[10] Iribarren SJ, et al. What is the economic evidence for mHealth? A systematic review of economic evaluations of mHealth solutions. PLoS One 2017;12(2):e0170581.
[11] Levine R, Corbacio A, Konopka S, Saya U, Gilmartin C, Paradis J, Haas S. mHealth compendium. Arlington, VA: African Strategies for health, management sciences for healthvol. 5; 2015.
[12] Chang LW, et al. Perceptions and acceptability of mHealth interventions for improving patient care at a community-based HIV/AIDS clinic in Uganda: a mixed methods study. AIDS Care 2013;25(7):874−80.

[13] Bavafa H, Hitt LM, Terwiesch C. The impact of e-visits on visit frequencies and patient health: evidence from primary care. Manag Sci 2018;64(12):5461−80.

[14] Nussbaum R, et al. Systematic review of mobile health applications in rehabilitation. Arch Phys Med Rehabil 2019;100(1):115−27.

[15] Georgsson M, Staggers N. Patients' perceptions and experiences of a mHealth diabetes self-management system. Comput Inform Nurs 2017;35(3):122−30.

[16] Weinhold I, Gastaldi L. From shared decision making to patient engagement in health care processes: the role of digital technologies. In: Challenges and opportunities in health care management; 2015. p. 185−96.

[17] Barello S, et al. eHealth for patient engagement: a systematic review. Front Psychol 2016;6:2013.

[18] Gupta A, Sao D. The constitutionality of current legal barriers to telemedicine in the United States: analysis and future directions of its relationship to national and international health care reform. Health Matrix 2011;21(2):385−442.

[19] Garattini C, et al. Big data analytics, infectious diseases and associated ithical impacts. Philos Technol 2019;32(1):69−85.

[20] Segura Anaya LH, et al. Ethical implications of user perceptions of wearable devices. Sci Eng Ethics 2018;24(1):1−28.

[21] Wiesner M, et al. Technology adoption, motivational aspects, and privacy concerns of wearables in the German running community: field study. JMIR Mhealth Uhealth 2018;6(12):e201.

[22] Zhou LA-O, et al. Barriers to and facilitators of the use of mobile health apps from a security perspective: mixed-methods study. JMIR Mhealth Uhealth 2019;7(4):e11223.

[23] Lucivero F, Jongsma KR. A mobile revolution for healthcare? Setting the agenda for bioethics. J Med Ethics 2018;44(10):685−9.

[24] Chee G, et al. Why differentiating between health system support and health system strengthening is needed. Int J Health Plann Manag 2013;28(1):85−94.

[25] Meskó B, et al. Digital health is a cultural transformation of traditional healthcare. mHealth 2017;3. 38−38.

[26] Blumenthal D. Doctors in a wired world: can professionalism survive connectivity? Milbank Q 2002;80(3):525−46.

[27] Rufo RZ. Use of change management theories in gaining acceptance of telemedicine technology. Crit Care Nurs Q 2012;35(4):322−7.

[28] Gaggioli A, et al. A telemedicine survey among Milan doctors. J Telemed Telecare 2005;11:29−34.

[29] Chambers DA, Glasgow RE, Stange KC. The dynamic sustainability framework: addressing the paradox of sustainment amid ongoing change. Implement Sci 2013;8(117):1−11.

[30] Scheirer MA, Dearing JW. An agenda for research on the sustainability of public health programs. Am J Public Health 2011;101(11):2059−67.

[31] Asiimwe C, et al. Use of an innovative, affordable, and open-source short message service-based tool to monitor malaria in remote areas of Uganda. Am J Trop Med Hyg 2011;85(1):26−33.

[32] Runkle D. The mHealth revolution. SciTech Lawyer 2013;9(3):24−5.

[33] Scott RE, Mars M. Principles and framework for eHealth strategy development. J Med Internet Res 2013;15(7). e155−e155.

[34] Alvarado MM, et al. Barriers to remote health interventions for type 2 diabetes: a systematic review and proposed classification scheme. J Med Internet Res 2017;19(2):e28.

[35] Muhambe TM, Ochieng DO, Wagacha PW. Proposing parameters for evaluating sustainability of mHealth systems in developing countries. Int J Comput Technol 2018;17(1):7153−62.

[36] Radhakrishnan K, et al. Barriers and facilitators for sustainability of tele-homecare programs: a systematic review. Health Serv Res 2016;51(1):48−75.

[37] Taherdoost H. A review of technology acceptance and adoption models and theories. Procedia Manuf 2018;22:960−7.

[38] Luna D, et al. Health informatics in developing countries: going beyond pilot practices to sustainable implementations: a review of the current challenges. Healthc Inform Res 2014;20(1):3−10.

CHAPTER 11

Digital health regulatory and policy considerations

Robert Jarrin, JD [1,2], Kapil Parakh, MD, MPH, PhD [3]

[1]Adjunct Assistant Professor, Department of Emergency Medicine, George Washington University School of Medicine and Health Sciences, Washington, DC, United States; [2]Adjunct Assistant Professor, School of Medicine - Department of Biochemistry and Molecular & Cellular Biology, Georgetown University Medical center, Washington, DC, United States; [3]Adjunct Assistant Professor, Yale University School of Medicine, New Haven, CT, United States

Contents

1. Learning objectives

The objectives of this chapter are to familiarize the reader with the role of government policies and regulations specific to healthcare services and medical products.

Digital Health
ISBN 978-0-12-820077-3
https://doi.org/10.1016/B978-0-12-820077-3.00011-0

2. Flow and rationale of the chapter structure

The chapter is intended to flow sequentially. The reader will be able to understand the general construct of how government works, with deeper insights on the most critical agencies that impact the broader areas of digital health.

3. Fundamentals of government and healthcare

Healthcare is one of the largest industrial sectors of the global economy, accounting for approximately $10 trillion dollars in spend every year.[1] Healthcare services are a complex interrelated web of businesses, vendors, insurers, providers, organizations, and most importantly, patients. Healthcare is highly regulated.

Regulations in the healthcare space are typically formulated from decades old frameworks, often ill-suited for the rapid advancement of technological innovation and growth. Digital health has become a disruptive force currently changing significant aspects of the business of healthcare. The pace of innovation has to be balanced with minimizing risk for patients. This chapter will briefly introduce the reader to how the US system of government is organized, how laws are created, how rules are derived from law, and how governmental agencies implement those laws, draft regulations, and set policy. This chapter will specifically highlight several agencies within the US Department of Health and Human Services (HHS) that affect the development, reimbursement, and ultimately the adoption of digital health.

4. How the US federal government is organized

There are three branches to the federal government—the legislative, executive, and judiciary. The legislative branch is authorized to make laws (the "Congress" consists of the US House of Representatives and the US Senate, although most people refer to the House simply as "Congress"). The executive branch implements laws (the executive branch includes the president, vice president, cabinet officials, and most federal agencies). The judiciary interprets the laws of the land (it includes the Supreme Court and lower federal courts).

---

[1] Deloitte: https://www2.deloitte.com/us/en/pages/life-sciences-and-health-care/articles/us-and-global-health-care-industry-trends-outlook.html.

All three branches are vital to the business of healthcare and influence each other through a system of checks and balances.[2] The US healthcare system is comprised of many private participants (vendors, insurers, providers, etc.) as well as public health agencies that range between federal, state, and local governments. The federal government provides states with funding to administer health programs. States in turn may distribute those monies to local health departments. States, through a system that mirrors the federal system with legislators and executives (governors), draft regulations and policies whereby local health departments implement healthcare services. For purposes of this chapter, we focus on the federal government and the agencies that implement and administer federal laws that regulate specific aspects of healthcare.

5. Health and human services law, regulations, and policy

Federal agencies fall under the jurisdictional authority of the President of the United States. Specific to healthcare, there are several federal agencies that administer rules and regulations and implement policy. It is important to understand that there are hundreds of federal agencies and commissions under Executive Departments, of which 15 are considered part of the President's "cabinet." The US Department of HHS is one such executive department. The mission of HHS is to enhance and protect the health and well-being of all Americans. It does so by providing health and human services and fostering advances in medicine, public health, and social services. HHS is administered by the Secretary of HSS, who is appointed by the President with the advice and consent of the Senate.

Under HHS, numerous programs are administered across its operating divisions, which include the Food and Drug Administration (FDA), the Centers for Medicare and Medicaid Services (CMS), the Office of the National Coordinator (ONC) for Health Information Technology, the National Institutes of Health (NIH), and the Centers for Disease Control and Prevention, among other agencies and offices. Each agency undertakes its specific authority to implement the programs and social services that it has been directed to manage by a Congressional Act—a Bill—once signed into law by the President. Federal laws "preempt" or nullify any state laws that conflict with them. Laws are statutes that serve as predetermined parameters for federal agencies to operate under in order to implement,

[2] https://www.usa.gov/branches-of-government.

interpret, and enforce program regulations, rules, and policies. The process of rule implementation or "rulemaking" is a vital mechanism of the federal participatory process. When developing regulations, the corresponding agency sets forth public notice asking if a regulation is needed, what it may affect, those who are affected, timelines for implementation, and all other associated information. Every regulation is somewhat unique in its scope and development, but generally follow a preset recipe: the issuing Agency publishes a "Notice of Proposed Rulemaking" or NPRM in the Federal Register so any member of the public or interested party may provide comments to the Agency; the Agency then considers the comments and may revise the proposed regulation; the Agency issues a final rule which is then codified in the Code of Federal Regulations (CFR), which serves as the record for all federal regulations created by the US Government. Once a regulation is finalized, the issuing Agency may add color to rules and regulations by issuing nonbinding policy that acts as guidance to further instruct interested parties on the government's desired outcomes.

6. Government and its impact on digital health

There's little question as to whether mobility, data, and information technology have served to disrupt the traditional culture of healthcare. Those disruptions, however, have created an environment of regulatory ambiguity. Is a health app downloaded from Google Play or the Apple App Store a medical device? Who should pay for the app—the patient, insurer, or the government? Can a medical professional using the app be reimbursed for spending time interpreting the information from the app? Does it matter if the medical service is performed by a physician, a qualified healthcare professional, or clinical staff? If the app happens to have a communication's function (either text or phone call), are those interactions between a healthcare provider and their patient protected under privacy laws or other regulations? What if the app is a medical device, can it work with unregulated devices and services such as an EHR? Many questions related to digital health and regulations, abound.

The evolving area of digital health does not fit neatly into the existing framework of policy, regulations, and laws governing the traditional healthcare space. A major challenge is that there are particularly broad definitions for things like what is a medical device, coverage and payment, and privacy and security. This presents a challenge when considering distributed interoperable and porous health IT systems that seek to liquefy healthcare data for myriad purposes beyond brick and mortar healthcare settings.

With the growth in popularity of fitness trackers, patient portals, and associated apps, consumers have begun to rely upon and use digital health products to better manage, track, and communicate their health status to physicians, peer support networks, and loved one's. The use of technologies such as smartphones, social media, and the availability of high-speed mobile broadband networks have on the one hand indelibly changed the reach of healthcare, but on the other, irreversibly complicating a somewhat predictable regulatory area.

6.1 Food and Drug Administration—clarifying digital health

The FDA has undergone tremendous change in how it approaches digital health. FDA is responsible for the oversight of more than $2.5 trillion in food, medical products, and tobacco.[3] According to FDA, products regulated by the Agency account for about 20 cents of every dollar spent by US consumers.[4] Part of this includes the oversight of over 6000 different medical device product categories.[5] Medical devices include obvious things like pacemakers, blood glucose monitors, and surgical instruments. But what about mobile medical apps, home health hubs, patient decision support software, or general health and wellness products such as fitness trackers?

The most basic question, "What is a medical device?" is defined by Section 201(h) of the Federal Food, Drug, and Cosmetic (FD&C) Act, which states

"an instrument, apparatus, implement, machine, contrivance, implant, in vitro reagent, or other similar or related article, including a component part or accessory which is: recognized in the official National Formulary, or the United States Pharmacopoeia, or any supplement to them, intended for use in the diagnosis of disease or other conditions, or in the cure, mitigation, treatment, or prevention of disease, in man or other animals, or intended to affect the structure or any function of the body of man or other animals, and which does not achieve its primary intended purposes through chemical action within or on the body of man or other animals and which is not dependent upon being metabolized for the achievement of any of its primary intended purposes."[6]

[3] https://www.fda.gov/about-fda/fda-basics/fact-sheet-fda-glance.
[4] Id.
[5] Id.
[6] Federal Food, Drug, and Cosmetic Act (1938), P.L. 75–717.

A key to understanding the section is understanding the context by which it is been shaped over the years. At various points since its original draft, Congress (prompted by historical events) has added to or amended the definition. Take for instance, how the Medical Device Amendments of 1976 followed a particularly damaging US Senate finding that faulty medical devices had caused over 10,000 injuries, including 731 deaths. Congress eventually moved to apply safety and effectiveness safeguards to new devices through "General Controls," which now represent the basic provisions (i.e., authorities) that provide FDA with the ability to regulate devices and ensure their safety and effectiveness. The General Controls in the Amendments apply to all medical devices and include provisions related to the adulteration; labeling or misbranding; device registration and listing; premarket notification; banned devices; notification, including repair, replacement, or refund; records and reports; restricted devices; and good manufacturing practices.[7]

Thus, the simple question of whether a product is a medical device starts with how the product is labeled or if it is used in a manner that meets the definition. If so, it will be designated and regulated as a medical device that is subject to FDA's laws and regulations before it is offered for sale or use in the United States. Many digital health solutions fall outside the scope of FDA's regulation since the FDA does not consider software that focuses on health promotion as a medical device. This approach stems from Section 3060 of the 21st Century Cures Act that passed in 2016 and has been further clarified in guidance documents issued by the FDA. When creating a digital health solution, it is not uncommon to choose an approach that does not create a medical device to reduce regulatory and compliance burden. However, some seek FDA clearance as an opportunity to diagnose or treat disease. Others see regulatory clearance as a way to differentiate their product in the marketplace and build a proverbial moat against competition. Of note, regulatory clearance is not needed for doctors to recommend a product but in some cases for insurers to pay for it. Given the complexity of this space, it is important to work seek advice and counsel from experts prior to making decisions with legal and regulatory consequences.

If the product is a medical device, then its risk level will determine the premarket requirements. Medical devices are classified into three medical

---

[7] https://www.fda.gov/medical-devices/regulatory-controls/general-controls-medical-devices.

device risk classifications: Class I, Class II, and Class III. This classification will help determine if mere listing (Class I or low risk) will be sufficient prior to marketing or whether a premarket notification (Class II or moderate risk) or premarket approval (PMA) (Class III high risk) is required. Class I are low-risk devices that typically do not require premarket notification prior to marketing. Class II are moderate-risk devices that require a 510(k) premarket notification submission to FDA (aptly named after the section of the Act) to demonstrate that the device is safe and efficacious and substantially equivalent to a "predicate device" that has already been legally marketed. If there is no predicate device and this is the first of its type, a de novo process will have to be followed. Class III are the high-risk medical devices that require a PMA. A PMA signifies FDA's most stringent procedural scientific and regulatory review to evaluate the safety and effectiveness of high-risk medical devices.

Devices may also fall under a regulatory mechanism used by FDA, where the specific product may meet the definition of a medical device, but because that particular device poses a lower risk to the public, FDA exercises "enforcement discretion" over that device—meaning the agency does not enforce regulatory requirements under the FD&C Act. Manufacturers may still choose to follow regulatory obligations including general controls as the product continues to be a medical device; however, FDA chooses not to require those regulatory obligations, thus doing so is up to the manufacturer.

All medical devices regardless of classification are subject to "General Controls," which include establishment registration (21 CFR Part 807.20) of the type of company and its involvement (i.e., manufacturers, specification developers, distributors, repackages, relabelers), medical device listing with FDA of the devices to be marketed, adhering to a quality management system (QMS) (21 CFR Part 820) through current good manufacturing practices (unless exempted), labeling and in some cases special labeling in accordance with labeling regulations (21 CFR Part 801 or 809), and may also include mandatory performance standards, and postmarket surveillance.

As we can see, a digital health product can be used to detect a parameter or intervene. If the product is being used to detect something and provide a diagnosis, it may be considered a medical device and fall into the regulated

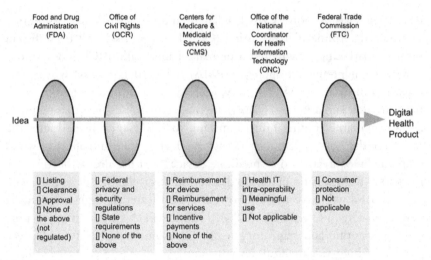

Figure 11.1 Summary of some of the regulatory considerations that are necessary to go from an idea to a product. It is important to understand why they exist and how they protect the public.

space. On the other hand, if the product is displaying information for educational purposes, it may be unregulated. In the case of intervention, if the product is providing treatment for a medical condition, it may be regulated. If it is helping promote wellness, then it may be unregulated. This framework is outlined in Fig. 11.2. Understanding where a product falls is important to determine the kinds of regulation it is subject to and the data needed to prove safety and efficacy.

6.2 Office for Civil Rights—overseer of healthcare privacy and security

Any digital health solution will, by definition, deal with electronic data. As a result, those solutions will most likely need to comply with health information privacy requirements. The Office for Civil Rights (OCR) enforces federal civil rights laws, conscience, and religious freedom laws, the Health Insurance Portability and Accountability Act (HIPAA) Privacy, Security, and Breach Notification Rules, and the Patient Safety Act and Rule, which together protect the fundamental rights of nondiscrimination, conscience, religious freedom, and health information privacy.[8]

---

[8] https://www.hhs.gov/ocr/about-us/index.html.

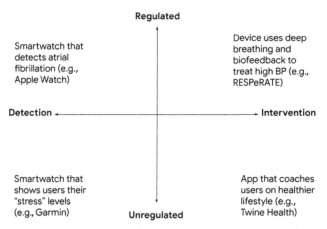

Figure 11.2 The spectrum of digital health products.

HIPAA is the prevailing federal statute governing privacy and security related to healthcare in the United States. It was signed into law in 1996, at a time when information technology was mostly wired (mobile phones were barely digital) and healthcare data largely took form in paper files stuffed into manila folders. HIPAA to this day provides security provisions and data privacy to keep medical information safe. The US Department of HHS publishes the HIPAA Privacy Rule and the HIPAA Security Rule. The Privacy Rule, or "Standards for Privacy of Individually Identifiable Health Information," establishes national standards to protect certain health information that is held by those who are considered "covered entities." Patients have rights with respect to individually identifiable information.[9] The "Security Standards for the Protection of Electronic Protected Health Information" (the Security Rule) establishes national standards for protecting certain health information that is held or transferred in electronic

[9] See https://www.hhs.gov/hipaa/for-professionals/covered-entities/index.html. A covered entity is a healthcare provider, a health plan, or a healthcare clearinghouse. Any individual or entity that meets the definition of a covered entity under HIPAA must comply with requirements to protect privacy and security of health information. If a covered entity engages another entity to assist with healthcare activities and business functions that involve protected health information, the new entity may be considered a "business associate." Business associate arrangements are governed through agreements that require the business associate to comply with HIPAA obligations to protect the privacy and security of protected health information. Business associates are directly liable for compliance with the HIPAA Rules. If an entity does not meet the definition of a covered entity or business associate, they do not have to comply with HIPAA Rules.

form. The Security Rule requires administrative, physical, and technical safeguards to ensure that confidentiality, integrity, and security of electronic protected health information. It is important to understand the different data types and protections accompanying them.

Personally Identifiable Information (PII) as defined in the Office of Management and Budget Memorandum M-07-1616 refers to information that can be used to trace an individual's identity, either alone or when combined with other information.[10] Some data (such as a social security number) are always considered PII, but other data require a case-by-case assessment of the specific risk that an individual can be identified.

Protected Health Information (PHI) is information that can be linked to an individual and generated by a covered entity (or its business associates) that relates to the health status of an individual (past, present, or future), and the provision of healthcare or payment for health services. The Security Rule, as outlined above, operationalizes the protections in the Privacy Rule with technical and nontechnical safeguards. Organizations called "covered entities," which include health plans, healthcare clearinghouses, and healthcare providers that electronically transmit any health information in connection with transactions for which HHS has adopted standards, must put in place protections to secure individuals' electronic protected health information. Electronic PHI has to meet a set of privacy and security standards as outlined in the HIPAA Privacy Rule and the HIPAA Security Rule.[11,12] While the specific approaches are beyond the scope of this chapter, it is important to take measures to protect PII and PHI when creating a digital health solution. In addition to the federal requirements, there are state privacy requirements to comply with. While these measures are mandated by law, they are also important to users. Research suggests that a lack of security features and privacy concerns are barriers to adoption of digital health solutions.[13]

[10] https://www.gsa.gov/reference/gsa-privacy-program/rules-and-policies-protecting-pii-privacy-act.
[11] https://www.hhs.gov/ocr/privacy/hipaa/administrative/privacyrule/index.html.
[12] https://www.hhs.gov/hipaa/for-professionals/security/index.html.
[13] https://www.ncbi.nlm.nih.gov/pubmed/30990458.

6.3 Centers for medicare and medicaid services—reimbursement of digital medical services

Although Medicare is composed of several benefit categories with innumerable programs, none exclusively rely, incentivize, or mandate the utilization of digital health as part of the provision of medical services. That does not mean providers cannot offer digital health services or implement digital technologies in their practice—they just would not get paid for it by Medicare. For purposes of this chapter, we will focus on physician and qualified healthcare professional services, as covered and paid in Medicare Part B. CMS provides health coverage to more than 100 million people through Medicare, Medicaid, the Children's Health Insurance Program, and the Health Insurance Marketplace. The annual budget for CMS as of 2019 topped $1.12 trillion dollars.[14] Despite the amount spent by CMS, until recently, very little was spent on digital health technologies or general remote physiologic monitoring (RPM) services rendered through digital medical modalities. Moreover, most digital health services were incorrectly labeled "telehealth" and relegated to Section 223 of the Social Security Act, which had been added to the Medicare, Medicaid, and SCHIP Benefits Improvement and Protection Act (BIPA) of 2000.

Since the inception of telehealth services in the year 2000 through BIPA, payment for telehealth has been notoriously low. In the 2001 Physician Fee Schedule, final rule, CMS promulgated 42 CFR 410.78, the regulation that establishes billing rules for telehealth under a CMS definition that "audio and video equipment permitting two-way, real-time interactive communication" between a patient and physician or practitioner was an interactive "telecommunications system." It also specified that telephones, facsimile machines, and electronic mail systems did not meet the definition of interactive telecommunications system, which were limited to live voice and video communications.

Perspective is key here because "modern" telecommunications in the year 2000 were not quite sophisticated. In fact, at the time there were more active landline telephones than mobile phones in the United States, mobile networks were capped at 2G services with barely any data throughput, and the world's first camera phone was introduced in the year 2000 with a maximum resolution of 0.11 megapixels and a 256 color display. In other words, technology lacked processing capabilities, there was no ubiquity, and

[14] HHS Budget in Brief: https://www.hhs.gov/sites/default/files/fy-2019-budget-in-brief. pdf.

high-speed data did not exist. 42 CFR 410.78 instructs that (1) telehealth technology must be synchronous (i.e., no store and forward unless it was accomplished through a federal demonstration project in Alaska or Hawaii); (2) the beneficiary must present in a health professional shortage area, or not in a metropolitan statistical area, or in an originating site as stipulated by CMS (skilled nursing facility, hospital, etc.); (3) the distant site practitioner must be a doctor, nurse, or stipulated medical professional; and (4) the service must be on the approved list of telehealth services (including psychotherapy, pharmacologic management, nutrition therapy, smoking cessation, transitional care management, or end-stage renal disease related service, to name a few).

Modern remote patient monitoring technologies typically do not offer or rely upon live voice or video. In fact, most remote monitoring of physiologic data is performed passively and taken from a patient through medical devices or body worn sensors that report readings over minutes, hours, and days. For whatever its reasons, CMS chose to strictly interpret the term "telecommunication system," under telehealth services and that decision significantly restricted the availability of telehealth and erroneously, all digital services seemed to be placed under this inaccurate moniker. These limitations have served as a hindrance to the adoption by providers of digital health solutions. In 2016, the Medicare budget (not including Medicaid, CHIP, and other CMS monies) was approximately $588 billion, while the level of telehealth reimbursement was $28.7 million.

It was not until the draft 2018 Physician Fee Schedule was delivered in July 17, 2017, that CMS distinguished general remote patient monitoring services from telehealth. In that notice of proposed rulemaking, CMS made a policy declaration on Remote Patient Monitoring.[15] CMS sought comments on whether to make separate payments for CPT codes that describe remote patient monitoring, while reminding the public that RPM are generally not "considered Medicare telehealth services as defined under Section 1834(m) of the Act." CMS stipulated that RPM services involve the interpretation of medical information without a direct interaction between the practitioner and beneficiary, and as such, "they are paid under the same conditions as in-person physicians' services with no additional requirements regarding permissible originating sites or use of the telehealth place of service code."

---

[15] Centers for Medicare and Medicaid Services, Physician Fee Schedule, and other revisions to Part B for CY 2018. July 21, 2017. https://www.govinfo.gov/content/pkg/FR-2017-07-21/pdf/2017-14639.pdf.

Medical Nomenclature—Coding is a vital part of the healthcare ecosystem. Without codes, there is no way to qualify or measure services, equipment, and procedures. No codes equate to no coverage, no payment, and no reimbursement. CMS relies heavily on various code sets that function as classification systems used to standardize clinical procedures, professional services, equipment, devices, pharmacology, and other descriptive nomenclature that cover every aspect of healthcare. Medical terminology is created by several organizations including the American Medical Association (AMA), the World Health Organization (WHO), the CMS, and the National Library of Medicine, to name a few. The AMA in particular plays a crucial role in the creation, administration, and management of the Current Procedural Terminology (CPT) code set that it maintains and copyrights. CPT is the most widely used listing of descriptive medical terminology and identification codes used by every physician and qualified healthcare professional in the United States. CPT serves as the uniform language that is continually harmonized among CMS, physicians, and other healthcare stakeholders. The CPT Code Set establishes the standard language for coding medical services and procedures that help streamline reporting and billing.

In September of 2017, the CPT Editorial Panel created several new RPM codes that would subsequently be adopted for coverage and payment by CMS in the 2019 Physician Fee Schedule starting on January 1, 2019. CMS had finally made a substantial advance toward digital medical services. In several future proposed rules, CMS would go on to remind that remote patient monitoring services are not Medicare telehealth services and therefore not subject to telehealth statutory or regulatory restrictions. This important distinction of remote patient monitoring from telehealth is fundamental. It demonstrates how agency statutory authority can be shaped through rulemaking. In fact, one can argue that federal agencies have much to achieve without an act of congress.

6.4 Office of the National Coordinator for health information technology—defining interoperability

Serving as the key resource to the entire healthcare industry to support the nationwide exchange of health information is the ONC. By incentivizing the adoption of health information technologies including digital health, ONC coordinates nationwide efforts to implement and use advanced health information technology through the exchange of electronic health record (EHR) technologies and associated data. ONC was initially established in 2004 under the then President Bush to build an interoperable and secure nationwide health information system. It was not until the passage of the American Recovery and Reinvestment Act of 2009, enacted into law by President Obama on February 17, 2009, that ONC along with CMS was provided the funding to create, define, and implement an EHR incentive program. Under "Meaningful Use" (as it came to be known), eligible professionals, eligible hospitals, and critical access hospitals would need to attest to the adoption of certified EHR technology and certified EHR modules.[16] These participants would earn incentive payments along with continued participation in Medicare and Medicaid. As ONC developed the technical standards for data exchange, it also became one of the principal agencies to explore how patient-generated health data from a variety of digital health data sources may become part of the EHR through the use of application programming interfaces used by smartphones to download and upload personal health data.

6.5 Federal Trade Commission—protector of consumers and competition

Federal Trade Commission (FTC) protects consumers and competition by preventing anticompetitive, deceptive, and unfair business practices through law enforcement, advocacy, and education without unduly burdening legitimate business activity. Its mission is to protect consumers from unfair and deceptive practices in the marketplace, particularly in healthcare and with a focus on healthcare and health IT. As an example, the FTC fined the developers of a mobile health (mHealth) apps for allegedly failing to pay promised bonuses to users who met health and wellness goals, as well as several apps purporting to identify melanoma from a photograph of a mole.[17] The FTC is the only federal agency with both consumer protection and competition jurisdiction in broad sectors of the economy. The FTC has the authority to stop and prevent unfair anticompetitive business practices such as price fixing, group boycotts, and exclusionary contracts, which can reduce competition and lead to higher prices, impair quality, and stifle innovation. Agreements between competitors may lead to violations of Section 5 of the FTC Act that bans "unfair methods of competition" and "unfair or deceptive acts or practices."[18]

---

[16] https://www.healthit.gov/topic/about-onc.

[17] https://mhealthintelligence.com/news/ftc-targets-mobile-health-app-developer-for-deceptive-practices.

[18] https://www.ftc.gov/enforcement/anticompetitive-practices.

The FTC pursues vigorous and effective law enforcement; advances consumers' interests by sharing its expertise with federal and state legislatures and US and international government agencies; develops policy and research tools through hearings, workshops, and conferences; and creates practical and plain-language educational programs for consumers and businesses in a global marketplace with constantly changing technologies. In 2016, the FTC created a web-based interactive tool for developers of mHealth apps, designed to help understand what federal laws and regulations might apply to those products.[19] Through a series of questions, developers are given basic answer to questions of legal jurisdiction, how those laws may apply to their products, and basic things to consider when protecting consumers' privacy and data security. The FTC developed the tool in conjunction with ONC, FDA, and OCR. FTC's work is performed by the Bureaus of Consumer Protection, Competition, and Economics.[20]

6.6 National Institutes of Health—providing research and clinical evaluation to digital health

NIH seeks fundamental knowledge about the nature and behavior of living systems and the application of that knowledge to enhance health, lengthen life, and reduce illness and disability. Through a series of workshops, grants, and initiatives, the agency has sought to foster a fundamental understanding of digital health to serve as a basis for protecting and improving health. NIH develops, maintains, and renews scientific human and physical resources that will ensure the Nation's capability to prevent disease. Examples include mHealth research through the Office of Behavioral Social Sciences Research, digital health as part of the Million Person Precision Medicine Initiative, the Health ePeople Research Platform, and other NIH research grants. Over the years, this has included a number of workshops and associated efforts related to mHealth and digital medicine.[21] In addition, the NIH houses two programs called Small Business Innovation Research and Small Business Technology Transfer, which are a source of early stage capital for technology commercialization in the United States.[22] Through these and other efforts, NIH seeks to expand the digital health knowledge

---

[19] https://www.ftc.gov/news-events/press-releases/2016/04/ftc-releases-new-guidance-developers-mobile-health-apps.

[20] https://www.ftc.gov/about-ftc.

[21] https://www.nih.gov/about-nih/what-we-do/mission-goals.

[22] https://sbir.nih.gov/.

base in medical and associated sciences to enhance the Nation's economic well-being and ensure a continued high return on the public investment of digital health. In realizing these goals, the NIH provides leadership and direction to programs designed to improve the health of the Nation. These programs can serve as an important resource for digital health companies looking to translate emerging science into products.

7. International regulation of digital health

While this chapter has focused on the United States, it is important to recognize that every country has a different regulatory and legal framework. Many countries follow similar constructs as the United States, but there are important differences too. Developers of digital health products are reminded to check with the appropriate regulators prior to developing and commercializing their digital health solutions. International differences between regulators can be subtle.

For example, medical device regulations in the European Union follow risk classifications according to Class I, Class IIa and IIb, and Class III designations. A Class I device requires registration of the manufacturing organization with the Competent Authority (e.g., the European Medicines Agency or the Medicines and Healthcare products Regulatory Authority [MHRA] in the United Kingdom). The manufacturer is responsible for ensuring that the QMS processes are defined and working to ensure the product meets the Essential Requirements set out in the Medical Device Directive. A declaration of conformity is signed by the manufacturer to show it meets all requirements. A Class IIa or IIb device would need a Notified Body approval of the QMS (under ISO13485) prior to affixing a CE mark and marketing the device and would need to have the technical file and initial clinical evaluation approved by the notified body (if the device falls under a Class IIb or Class III).

Similar to FDA in the United States, if there is not enough existing data to prove substantial equivalence to an existing approved device, then clinical trials may be required. An ethics committee and the MHRA or other body have to approve the studies being done and the studies must be conducted under strict privacy and ethical standards. Once the MHRA approves a device, the manufacturer also has the responsibility of ensuring the device is safe through postmarket surveillance, complaint handling, documentation, and other processes. In order to harmonize broad aspects of medical device regulations, such as Software as a Medical Device, the

International Medical Device Regulatory Forum was established in February 2011. This voluntary collective of medical device regulators from around the world, including Europe, Asia, the Americas, and Africa, work to harmonize regulatory positions and concepts. Unfortunately, there are few equivalent organizations seeking to harmonize other aspects of healthcare law and regulations.

8. Conclusion

The US federal system allows the three separate branches of government to operate independently of each other, yet they are inextricably bound in a symbiotic relationship that can never be undone. This "separation of powers" prevents overwhelming concentration of power in any one branch and to protect the rights and liberties of citizens. We have discussed how the executive branch implements laws and shown how agencies fall under the jurisdictional authority of the President of the United States. We have looked deeper at how several federal agencies administer rules, regulations, and implement policy specific to healthcare. For the growing area of digital health, we have come a long way, yet much more needs to follow.

9. Description of pedagogical elements (e.g., case study, infographics required, key references)

Fig. 11.1 as above and attached separately.
   Fig. 11.2 as above and attached separately.
   References in footnotes.

Index

Printed in the United States
By Bookmasters